THE WAR ON SCIENCE

CHRIS TURNER

Muzzled
Scientists and
Wilful
Blindness in
Stephen Harper's
Canada

The
WAR
on
Science

GREYSTONE BOOKS
Vancouver/Berkeley

For the Turner4YYC campaign team

Greystone Books Ltd.
www.greystonebooks.com

Cataloguing data available from Library and Archives Canada
ISBN 978-1-77100-431-2 (pbk.)
ISBN 978-1-77100-432-9 (epub)

Editing by Nancy Flight
Copy editing by Lesley Cameron
Cover and text design by Jessica Sullivan
Printed and bound in Canada by Friesens
Distributed in the U.S. by Publishers Group West

We gratefully acknowledge the financial support
of the Canada Council for the Arts, the British Columbia Arts
Council, the Province of British Columbia through the Book
Publishing Tax Credit, and the Government of Canada through
the Canada Book Fund for our publishing activities.

Greystone Books is committed to reducing the
consumption of old-growth forests in the books it publishes.
This book is one step toward that goal.

CONTENTS

1

MARCH OF
THE LAB COATS

The View from the Street and the Lab
SPRING—SUMMER 2012

THE PROTEST march that snaked through the streets of Ottawa on the morning of July 10, 2012, was, in some respects, a standard affair. The marchers carried placards and chanted slogans, a roster of speakers made high-minded speeches, the police redirected traffic and kept a watchful eye. Under a bright blue sky, the protesters marched from the Ottawa Convention Centre past the Chateau Laurier to Parliament Hill, drawing the curious interest of the odd tourist or passerby, but mostly tromping down the streets of the capital with order, purpose and calm.

The only obvious signs that this was a protest unique in the history of Canadian public life were the crisp white clinical lab coats on dozens of the protesters and the geeky twist they gave to a familiar chant.

"What do we want?"

"Science!"

"When do we want it?"

"After peer review!"

A young woman carrying a scythe and wearing the black hooded robe of the Grim Reaper led the procession, followed by a clutch of pallbearers bearing a prop coffin aloft. The march had been billed "The Death of Evidence." It had been organized and was largely peopled by scientists—field researchers, lab rats, graduate students—and it was, as far as anyone marching was aware, the first time their ranks had ever assembled to stage a protest on Parliament Hill.

Scientists are by professional tradition and often by general disposition a cautious, reserved lot. They place the highest virtue on reasoned argument and cloistered study, proceeding from the core belief that scientific evidence, objectively gathered and impartially analyzed, must always trump opinion and argument and shouted slogan in the establishment of what is true and reasonable and which courses of action best serve the public interest. They conduct their public discourse as much as possible in the meticulous, technical language native to peer-reviewed scientific journals. That the scientists in Ottawa had taken their conversation to the streets, amplified it, reduced it to the crude exigencies of a placard's slogan—this all spoke to a catastrophic decline in the harmony of their usual dialogue with Canadian government.

In Canadian public life, there had for generations been a sort of implicit understanding between scientists and politicians, between those who gathered and analyzed the data and those who used the resulting studies and white papers and policy briefs and committee testimonials to enact legislation. It went roughly like this: federal law and public policy would always have recourse to the best available evidence. Any number of political persuasions and points of view could be represented in the public discourse—rabid socialists and staunch libertarians, rapacious capitalists and bleeding-heart liberals, Tories and Grits, Dippers and Greens—but scientific evidence existed outside this cacophonous arena of competing opinions. The parameters of the entire debate were established by observable, verifiable, peer-reviewable reality, not by political expediency or strategic advantage. Even if this evidence-based social contract was not always honoured in full, it had never been unilaterally negated. Politicians might elide inconvenient facts or omit problematic details in the name of short-term gain, but they weren't permitted to dismiss the scientific method itself as irrelevant to the formulation of policy. At some point you had to acknowledge the basic facts of the situation. Didn't you?

Since Stephen Harper's Conservatives had first formed a government in 2006, the pact between evidence and policy had eroded and crumbled and then finally given way at some fundamental level—the one that sent scientists marching in their lab coats on Parliament Hill. The process had been slow and sporadic at first—esoteric programs cut here and there, experts and their studies forced into the custody of media handlers, their conclusions massaged to corroborate talking points dictated by the Prime Minister's Office. The campaign intensified in fits and starts through the minority years, with rumblings about discounted evidence and silenced scientists accompanying the elimination of the Office of National Science Adviser, the cancellation of the long-form census, and the tabling of a sweeping crime bill that went against decades of research.

In the first year of Harper's majority, though, the scientific community's concerns turned quickly to outrage. Seemingly every other day through the spring of 2012, news broke

of another slashed budget or shuttered research facility as
the Conservatives rolled out Bill C-38, their omnibus budget
bill. Cloaked in the plain cloth of financial necessity, the bill
seemed intent on rewriting the whole contract between scientists and policy-makers.

"It was staggering," Dalhousie University biologist Jeffrey
Hutchings told me. "It seemed like every week there was something new happening. And it got to the point where as a Canadian scientific community, not only were we thinking what's
going to happen next, but there was so much, we didn't actually
know how to respond. And it really felt like a boxer who's just
been punched so many times, it's all you can do to stand up."

On July 10, Hutchings made his stand. He was in Ottawa
for a conference on evolutionary biology, an event co-sponsored by the Canadian Society for Ecology and Evolution.
Hutchings, the society's president, was a co-host, and there
was plenty for him to attend to inside the conference hall.
But the march's organizers had been in touch, encouraging
him to participate. It wasn't an easy decision. He was wary
of the planned Death of Evidence march, concerned that his
involvement might be interpreted as political advocacy or the
pursuit of personal gain. "I don't want to be advocating for
anything other than the communication of science," he said.
"So I was a little reluctant."

A chat with a journalist just a few days before the march
finally swayed him. On the appointed morning, Hutchings
left the conference centre and stepped out onto Colonel By
Drive in his crisp white lab coat, melting into the ranks of his
white-coated colleagues, all of them joined together in word-
less solidarity with the protest's simple assertion that scientific evidence was sacrosanct, that the final arbiters of truth
toiled not in the House of Commons but in the laboratory.

Phalanxed by placards, chants echoing in their ears, the nation's foremost evolutionary biologists began their march on Parliament Hill.

The placards themselves told a shorthand version of the story. One near the front of the ranks bore the march's semi-official motto: "NO SCIENCE/NO EVIDENCE/NO TRUTH/NO DEMOCRACY." Another put it more bluntly: "STOP HARPERS WAR ON KNOWLEDGE."

The marchers had first assembled outside the convention centre because a good many of them were, like Hutchings, in town for the conference. The First Joint Conference on Evolutionary Biology was one of those insular academic affairs whose symposium titles spoke in language guaranteed to exclude all but the fully initiated. *Influential symbionts: Master manipulators of adaptive host behavior. Genome evolution and speciation: "Next-generation" genomics of parallelism and convergence.* When scientists spoke to each other, this was their preferred lexicon—technical and data-driven, footnoted and impartial.

The protest's ranks had been filled out by busloads of instructors and grad students from McGill, Queen's and Waterloo, as well as a handful of more seasoned activists rallied by the Council of Canadians. But for most of the people following the Grim Reaper down Colonel By Drive that day, this was their first-ever protest march.

A petite, dark-haired young woman who had positioned herself to the side of the avenue watched anxiously as the marchers shuffled past, a megaphone clutched in her hand. This was Katie Gibbs, just weeks from completing her PhD in biology at the University of Ottawa. She was one of the few scientists present who had any real experience in the blunt craft of politics. Gibbs had been active in the federal Green Party for several years, and she was a co-organizer of the Death of Evidence march. She'd been there when her colleagues first proposed the idea of a protest on Parliament Hill over pints at an Ottawa pub a few weeks earlier, and they'd chosen her as lead organizer. Her colleagues, who had previously treated her outspoken political stance as a strange and potentially dangerous character flaw, were finally ready to concede that the Conservative government led by Stephen Harper had left them with no other avenue for dissenting dialogue. "It was interesting," Gibbs said, "to see their minds shift to the point where they seemed to realize that if we don't stand up for science then nobody else is going to."

As her peers strode past on the morning of the march, Gibbs craned her neck to try to find the end of their ranks. She'd been a nervous wreck about the turnout. If five hundred people came, at least she wouldn't be embarrassed. She was hoping for a thousand, but no one who had cut their political teeth on the Green Party fringe could be so foolhardy as to expect all their best-case scenarios to play out. As she watched, though, her anxiety turned to exuberance. For several amazing minutes, she couldn't find the end of the line. The marchers just kept coming.

There was a makeshift Dixieland jazz band at the front of the march alongside the pallbearers, bopping its way through "When the Saints Come Marching In." It lent a jaunty cadence to the marchers as they strolled past Gibbs—young and old, long-haired and bald-pated, in shorts and sandals or in button-down business casual. An older couple in matching Tilley hats, a young couple with one baby on the mother's hip and another in a stroller. Lab coats, black T-shirts, sundresses. Certain sections of the march, full of smiling faces and floppy summer hats and cameras, could have been mistaken for a sightseeing group expelled from some outsized tour bus. Gibbs hollered into her megaphone from time to time to instruct the neophyte marchers on how to stay in an orderly line. Someone had mounted a telescope on top of his bike helmet. One lab-coated woman held a sign up that read "WE ARE NOT RADICALS." Another sign said simply: "[CITATION NEEDED]." By the hundreds and hundreds, they just kept coming. The RCMP would eventually fix the estimated size of the crowd at two thousand, but Gibbs was sure it was larger than that by a substantial margin. And although it was far from a rowdy crowd, it was an exhilarated one.

Gibbs: "A lot of the people who participated really were scientists, and for a lot of them this was the first kind of activism or public campaign that they really had been a part of. I think for most of them this was the first time marching in the streets. So there is that kind of indescribable buzz that comes from doing something like that with a lot of other people."

One of the first marchers to reach the steps of Parliament's iconic Centre Block was Diane Orihel, a PhD student in aquatic biology at the University of Alberta. She was scheduled to address the crowd once it had fully assembled on the broad concrete expanse in front of her. Like Gibbs, she goggled at the size of her audience. "I was just overwhelmed with the number of people who came out to support us," Orihel recalled. "I remember standing on the steps of Parliament Hill and just watching the people flood the square, and they just kept coming and coming and coming. Soon the RCMP had to let people overflow onto the grass, because they had completely filled the entire area."

FOR ORIHEL, THE journey had begun six weeks earlier in bewilderment, fury and despair. She had arrived at her office at the Freshwater Institute in Winnipeg on the morning of

May 17 for a business-as-usual day. A colleague told her an emergency meeting of the whole institute staff was just about to start. "This can't be good news," the colleague said.

No scientist working on a federally funded project in the spring of 2012 could have been wholly complacent about their job security, especially if their field was in the environmental sciences. Bill C-38 had unleashed a broad frontal assault on the Canadian environmental science community. Tabled in the House of Commons six weeks earlier, the bill had triggered wave after wave of closures and "affected letters" (notices of potential or impending layoff) at research institutes, monitoring stations, and government labs across the country. The bill's "scale and scope," wrote the *National Post*'s Andrew Coyne, "is on a level not previously seen, or tolerated."

Here is Coyne's summary of the bill's extra-budgetary dimensions:

> It amends some 60 different acts, repeals half a dozen, and adds three more, including a completely rewritten Canadian Environmental Assessment Act. It ranges far beyond the traditional budget concerns of taxing and spending, making changes in policy across a number of fields. . . . The environmental chapters are the most extraordinary.

In an op-ed in the *Guelph Mercury*, Cynthia Bragg argued that "the federal government is taking a sledgehammer to environmental protection across Canada." In truth, it was more like a hundred vicious scalpels, and one of them had sliced away the Experimental Lakes Area (ELA), the array of fifty-eight small lakes in northern Ontario where Orihel and her Freshwater Institute colleagues conducted their research.

The news at the institute's emergency meeting was devastating. Orihel: "They all received affected letters, workplace suggestion letters, and were basically told to go out to the ELA and get their stuff out of the lakes, get the stuff out of the labs and pack it up, and no new research was to be done. They were also specifically warned that they were not allowed to communicate with the media or the public about the ELA."

The ELA is not a lab in any conventional sense but rather a sort of contained biosphere where experiments sometimes involve altering the fundamental biology and chemistry of a whole lake—or several lakes—for years at a stretch. It may be the most important freshwater research facility on the planet, and its researchers—co-founder David Schindler of the University of Alberta, in particular—had made discoveries of global import, uncovering the mechanisms by which acid rain poisons aquatic ecosystems and industrial phosphorus runoff damages freshwater chemistry. Given the scale and duration of the typical ELA experiment, the orders from Ottawa were akin to asking farmers to pack up their farms. (With a bumper crop in the ground. In the midst of a global famine.)

Because she was not a Freshwater Institute employee but a grad student, Orihel was one of the few ELA insiders who could respond freely. She soon found herself the de facto press officer for the ELA's defenders. She had never written a press release in her life, had to be told what the expected length was, where to place her contact information, how to put an old newsie's "-30-" at the end. "The next day I was so overwhelmed with media," she said. "And I realized that I had to be the public face of this, because all my colleagues were muzzled. I was an effective hub through which information could get out, because I have been working at ELA for ten years.

I have had close relationships with the current and former scientists for the last decade. I was perfectly aligned to be the public face and I had the least to lose, because I can't be fired."

The day after her debut as press officer, she established a grassroots organization, dubbed it the Coalition to Save ELA, and circulated a petition. By early June it had attracted 1,700 signatures, and Orihel flew to Ottawa to take the petition to Parliament. She organized four press conferences at the National Press Theatre and brought in scientists and opposition MPs to defend the ELA's track record. And she met a fellow grad student named Katie Gibbs, who had just started to organize a march on Parliament Hill.

FOR THE DEATH of Evidence march, Orihel wore a long gown in funereal black. Her head was half covered in a black scarf, and she gazed through dark sunglasses over the crowd stretching away from the Centre Block steps. She stood flanking the mock coffin, a couple of steps behind co-emcees Katie Gibbs and Scott Findlay. Findlay's inclusion had been Orihel's idea. She knew the protest needed at least one public face that was not a youthful student's, someone who couldn't be dismissed as a usual suspect. It was too easy for the media to write off the march as a summer lark by a bunch of students with nothing better to do. Findlay was a professor at the University of Ottawa, a working evolutionary biologist with a list of peer-reviewed publications to his name. He lent gravitas, made the soapbox safe for other speakers on the rally's docket: Jeff Hutchings; Arne Mooers, a biodiversity professor at Simon Fraser; Findlay's U of O colleague Vance Trudeau.

Katie Gibbs spoke first. She read a short obituary for evidence and introduced her fellow speakers as "eulogists." "We are here today to commemorate the untimely death of evidence in Canada," Gibbs announced. "After a long battle with the current federal government, evidence has suffered its final blow." The crowd responded with hoots and howls of "Shame!" The eulogists, using the stump speech rather than the professorial lecture as their model, kept their comments to a brisk two minutes or less. Vance Trudeau talked about the wilful blindness of the government's "anti-science" ideology. "The tendency to only use the data and evidence you like is the misuse of information for alternative purposes," he said. "This is known as propaganda." The next speaker was Arne Mooers, who worried about the country's ability to address the challenge of climate change without the best available data. "To deny evidence," he said, "is to live in a fantasy world."

Hutchings, despite his earlier reluctance to join the march, revealed a firebrand streak and a deft rhetorical touch. "Freedom of expression is no longer a right enjoyed by Canadian government scientists. These individuals paid by taxpayers to undertake research in support of society are not permitted to speak to Canadians unless they have ministerial permission to do so. When you inhibit the communication of science, you inhibit science. When you inhibit science, you inhibit the acquisition of knowledge. Government control over the ability of society to acquire knowledge has alarming precedents. An iron curtain is being drawn by government between science and society. Closed curtains, especially those made of iron, make for very dark rooms."

After University of Ottawa PhD student Adam Houben read a brief statement on behalf of his fellow students, Diane Orihel took the microphone. She kept her sunglasses on against the glare and delivered her speech from prepared notes in a careful, steady tone. She ran through the singular track record of the ELA, "a living laboratory unlike any other in the world," which had drawn the best scientists from around the globe to study the effects of phosphorus, mercury, nanosilver particles and much else on the lakes.

"Now on the verge of its untimely death," Orihel continued, "the Experimental Lakes are grappling with unresolved problems, such as how climate change alters lakes and how synthetic nanoparticles—put on our clothes to kill bacteria—may be a new threat. The lakes may have answered these questions in time to prevent great human and environmental suffering, as well as economic hardship. But, alas, the lakes are marked for death. Today we mourn the Experimental Lakes. But moreover, we mourn the blindfold of ignorance imposed upon our once-great country. With fear and trembling, we enter a two-plus-two-is-five universe."

The audience roared, and then the rally was over. And for all too many protests, that would have been the end of it. The marchers would have their memories, the media would file perfunctory reports, and the two-plus-two-is-five universe would spin on, oblivious. But instead of fading, the Death of Evidence march went supernova. Press coverage was substantial, prominent, engaged. The march made headlines coast to coast and beyond. The international wires picked up the story, and it found its way to the pages of foreign newspapers. The *Guardian* published a commentary on the Death of Evidence under the headline "Why Canada's scientists need our support." The piece was widely discussed and generated even more international coverage for the march.

In the days and weeks that followed, commentators began to identify parallels between doctors' protests against cuts to refugee health care and the heated opposition to Bill C-38 voiced by former Fisheries ministers and the quiet, orderly march of the lab coats on Parliament Hill. "Not only was the protest unprecedented, even extraordinary," wrote the *Toronto Star*'s Christopher Hume, "it struck at the dark heart of the New Canada, a nation more interested in hiding the truth than understanding it, exploiting resources than conserving them."

Just weeks after the protest, Prime Minister Stephen Harper told the press that proposed oil pipeline projects would be evaluated "on an independent basis scientifically, and not simply on political criteria." The emphasis in the statement seemed to be on *scientifically*; to the organizers of the Death of Evidence march, it sounded as if their rebuke just might have been heard even by the prime minister. What government, after all, would want to be seen as an enemy of science?

This, of course, was the march's intended message: that some topics should be beyond the reach of political point scoring, that data and research and reason and evidence are not arguments to be spun or opinions to be dismissed but the raw materials for establishing truth, the essential tools by which an enlightened government creates policy. "The Harper government," Orihel told me, "is strategically undermining our capacity to conduct science in this country and is systematically destroying not only the funding for the programs but also the research platforms for us to conduct the work. It is also systematically destroying the ability to use the science that we produce in order to move forward environmental protection. What's the point right now of conducting science that links, let's say, fish productivity with fish habitat, because the fish habitat provisions have been taken out of the Fisheries Act?"

What do we want? SCIENCE! *When do we want it?* AFTER PEER REVIEW!

At its core, this was not an argument to alter the compo-sition of Parliament, to change a specific policy, or even to restore funding to a particular program. The Death of Evi-dence march was a rallying cry for an idea that has not been particularly radical in the Western philosophical tradition since the 1700s—the idea that science is the final arbiter of truth. That such an assertion required street protest in the capital of a democracy in the twenty-first century speaks vol-umes about how drastically Canada has been altered under the government of Stephen Harper.

For as long as Canada has been a nation, Parliament Hill has been the scene of debate and protest, petty point-scoring and cronyism. It has seen more than its share of ignorance and delusion. It has never lacked for ill-informed policy-making or short-sighted argumentation. It was here, during a singularly grave crisis, that a prime minister famously sought the coun-sel of his deceased dog. Misreading science, ignoring the best available evidence, and letting ideology trump reason and faith trump the facts—the Hill's seen it all before. But a deeper breach of the public trust is at work in Stephen Harper's Can-ada. This is a government that does not simply ignore the best evidence but attempts to destroy the sources of it, a govern-ment that does not merely disregard the advice of experts but prohibits them from speaking about their work in public. It's one thing to dismiss good data in the name of political expe-diency and another thing entirely to wage war on the produc-ers of controversial data, to deny access to the best possible evidence not only to the current government but to all gov-ernments and every Canadian now and forever. Governments squabble and obfuscate and political parties trade in myth, rumour and specious argument as a matter of course, but only Harper's Conservatives have taken the fight so viciously to the nation's laboratories and field research stations. The lab coats marched on Parliament Hill not because this government was wrong but because it was so egregiously wrong about *what government is for.* The dispute was not with a single policy but about how policy is to be made. For as long as Canada has been a nation, science has stood outside the partisan fray, too often overlooked or downplayed but never, until now, attacked directly for partisan purposes. This is why the lab coats marched: the very foundations of scientific inquiry and the scientific method had been called into question.

So then: let 2012 be remembered as the year Canada's scientists marched in the streets of the capital to restore the basic principles of the Enlightenment to its government. The year, unbelievably, it came to that.

2

LANDSCAPE AT TWILIGHT

The View from Parliament Hill

SPRING–SUMMER 2012

THE DEATH of Evidence march was, in some sense, not the launch of a protest movement but rather the culmination of a protracted dispute, the nadir of a downward spiral. By 2012 the Canadian political landscape had already darkened ominously. For many months leading up to the march and for many months after, Stephen Harper's Canada was a landscape in twilight, a place where immediate political advantage and short-term economic gain stood alone on the highest ground, casting long shadows over hard data and reasoned analysis.

If Parliament was a dark and troubled scene, the wider world was darker still. The Canadian government's mounting disdain for the work of its scientists had not developed in isolation or against a backdrop of particularly calm, fair-weather skies. Science is an examination of the facts of the material world and the natural systems that govern it, and in 2012 those systems delivered clear signals of profound crisis—the same ones they had been broadcasting for many years.

"I believe there has been no worse year for the natural world in the past half-century," the *Guardian* columnist George Monbiot wrote at the end of 2012. The year was one of the ten warmest for the planet since record-keeping began in the 1800s—the fourth-warmest in Canada, the warmest ever in the United States. The news of unprecedented climatic chaos had already become almost commonplace, an expected series of oddities and devastations, an extreme reality TV series renewed annually. And 2012 added several wild new episodes.

That year alone, drought devastated the American Midwest and Southwest, western Russia, and great swaths of China. Australia sweltered under a catastrophic heat wave. The most ferocious storm in generations pounded the American Northeast, flooding the Jersey Shore and New York's outer boroughs like an animation in a disaster movie. In Canada, an overwarm March collided with a vicious April frost to wipe out 80 percent of Ontario's apple crop. As western Canada thawed, floods engulfed British Columbia's interior, even as the region's burgeoning army of mountain pine beetles renewed their relentless assault on mountain forests. By the end of the year, their voracious campaign had spread to more than one-fifth of the entire land area in British Columbia. A report by University of Toronto environmental scientist Holly Maness published in November showed not only that the forests had been killed by the warming climate that had allowed the beetles to thrive but that the dying trees were now contributing to further climate change, as the sun's rays struck the forest floor unimpeded by the cooling dew that used to blanket the trees' branches. As the Death of Evidence marchers mounted their protest on Parliament Hill, the Canadian Arctic was well into the most extreme sea-ice melting season in recorded history; by summer's end, a patch of ice nearly the size of India had dissolved into the ocean. The scope and pace of the thaw was well beyond the cautious predictions of the most recent report of the Intergovernmental Panel on Climate Change (IPCC). In Canada's Far North, the worst-case scenario had become reality in 2012.

Canada's geopolitical affairs have been no less chaotic in recent years. The Canadian government is now given Fossil of the Day awards at international climate negotiations as a matter of routine. Canada's track record on greenhouse gas emissions ranks dead last among G8 nations, and the Conference Board of Canada placed Canada fifteenth out of the world's seventeen richest nations in overall environmental stewardship. For climate activists around the world, the oil-extraction industry in northern Alberta has become a synonym for environmental destruction, turning an obscure pipeline project that the Conservative government had been certain was a "slam dunk" into a political lightning rod in Washington, D.C., as well as a rallying cry for protesters across North America and around the world. Another proposed pipeline across northern British Columbia has fuelled similar outcry in Canada, while First Nations across the country have launched their most widespread protest campaigns in a generation in response to a range of damaging environmental policies, funding cuts, and stalled treaty negotiations. Meanwhile, an *E. coli* outbreak at an Albertan meat-processing plant led to the largest beef recall in Canadian history and brought the instability of the country's food security regime into high relief. The year began with the humanitarian crisis in the Attawapiskat First Nation only just resolved—at least provisionally—and ended with the launch of "Idle No More," a vociferous nationwide First Nations protest movement. In the months in between, 116 First Nations communities—almost one in five—toiled under "boil water" advisories due to the abysmal state of their water supply.

The natural world broadcasts messages of crisis as never before, unprecedented in their urgency and frequency. Scientists, political leaders, activists, and civil society groups across Canada and around the globe speak with mounting alarm about Canada's wayward drift on environmental stewardship and climate action. And in response, the Canadian government has instigated a systematic, sustained campaign—without precedent and without recourse to best evidence—to cripple its ability to detect and respond to such crises, to monitor environmental damage and deal with disasters, even to conduct and communicate basic science in the public interest. In the face of imminent and emergent disaster, the government has opted for wilful blindness. Twilight has come to the Canadian landscape—an artificial twilight, manufactured in Ottawa by deliberate policy measures—and the Enlightenment itself has fallen into shadow.

In retrospect, the government's decision to eliminate the long-form census in 2010 was the bellwether of the broader campaign to come. It was arbitrary and illogical, a deliberate choice to limit the amount of data the government keeps at hand to fulfill its most basic duties and employs daily to make its most fundamental decisions. Without the long-form census, nearly every aspect of responsible governance is impeded. Health officials lose critical information about where their services are most needed. Labour shortfalls are harder to detect, oversights in the training of skilled workers are less easily identified, the Consumer Price Index—a vital piece of data for setting monetary policy—is rendered less precise. And on and on, from department to department and file to file— education, transportation, economic policy, justice, housing, pensions—and from one level of government to another and one jurisdiction to the next and the next.

"No country can be among the league of civilized societies without intelligent policy development," wrote former chief statistician Munir Sheikh, who resigned from his job at Statistics Canada rather than work in the gathering darkness. "And intelligent policy development is not possible without good data." A website called Datalibre.ca has tracked opposition to the cancellation of the census; at the most recent count, the declared opponents of the measure numbered 488, occupying a swath of civil society broad enough to include Alberta Health Services, the Canadian Arab Foundation, the Canadian Institute of Actuaries, the Ontario Chamber of Commerce, and the United Way, as well as half a dozen provincial and territorial governments and several dozen municipalities. When the first results from the new, voluntary National Household Survey (NHS) were released in May 2013, nearly every report included a disclaimer about the unreliability of the data. "The NHS estimates," one of the most prominent warnings read, "are derived from a voluntary survey and are therefore subject to potentially higher non-response error than those derived from the 2006 census long form."

The cancellation of the long-form census would prove to be the template for the Conservative government's approach to other matters involving the interpretation of hard data and scientific evidence. The same basic principles—centralized decision-making, arbitrary cuts, a stubborn refusal to consider opposing points of view—were soon evident far beyond the quiet counting offices of StatsCan. The Canadian government under Stephen Harper aims to speak with a single voice and hear nothing but the harmonious echo of its own wisdom. In order to eliminate the troublesome dissonance of inconsistent messages, government communications across all departments have grown steadily more centralized—first within departmental media affairs offices and then increasingly within the Prime Minister's Office itself. Environment Canada—among the most tightly controlled departments under the Conservative government—outlined its new communications protocol for dealing with media inquiries in a 2007 statement: "Just as we have one department we should have one voice. Interviews sometimes present surprises to ministers and senior management. Media relations will work with staff on how best to deal with the call. This should include asking the programme expert to respond with approved lines."

Asking the programme expert to respond with approved lines—a rather soft-edged way of saying that experts, including scientists working in a tradition of open inquiry and professional debate that dates back to the days of Newton, should report not what they have learned in the public interest but what the current occupants of Parliament's majority benches tell them to say. Science has long been treated by the Canadian government as a vital tool for informing policy; under Stephen Harper, the relationship has been inverted. Policy now determines what scientific evidence will be used to inform government and which scientists will be permitted to inform the public.

This innocuously worded protocol provoked a sea change in Canadian journalism. The media's small cohort of dedicated science-beat journalists had long enjoyed the same collegial relationship with government-employed scientists as they did (and still do) with university researchers. Background detail or an informative quote was rarely more than a quick phone call away. But now government scientists in every field have been obliged to pass on even the most mundane queries to their affiliated media officers. On topics deemed sensitive or controversial—which is to say, on topics where scientific evidence does not unequivocally support government policy— media reps have sat in on interviews and even banned government scientists from talking to the press about their work. When Kristina Miller of the Pacific Biological Station published a groundbreaking report on collapsing Pacific salmon stocks in the prestigious international journal *Science* in early 2011, for example, her minders from Fisheries and Oceans Canada's media relations department squelched a draft press release and refused numerous requests for interviews from major national and international media outlets. The point is worth underscoring: a scientist whose research was funded by Canadian taxpayers was forbidden from telling the Canadian public or anyone else what she had learned about a matter of vital national interest and global import.

Later that year, Mike De Souza of Postmedia News endured weeks of bureaucratic runaround to secure a phone interview with Environment Canada's David Tarasick, who had just published a report documenting his discovery of a substantial hole in the ozone layer over the Canadian Arctic. When De Souza asked about the delay, the response came not from Tarasick but from an Environment Canada media staffer sitting in on the phone call, who dismissed it as a "moot point." "I'm available when Media Relations says I'm available," Tarasick told De Souza.

Perhaps the starkest example of the government's top-down approach to science communications came at the International Polar Year conference, hosted in Montreal in April 2012.

This was, as already noted, just before the start of the most extensive sea-ice melt ever recorded—a phenomenon that would soon be described by scientists and journalists around the world as a catastrophe. Every one of the Environment Canada scientists who presented or sat on a panel at the conference was accompanied by a media handler. The handlers monitored the scientists' presentations and redirected follow-up queries to themselves. Never before had Environment Canada researchers required media escorts at major conferences or anywhere else, and no other country's scientists arrived with chaperones. A department spokesperson told Postmedia's Margaret Munro this was "standard practice"; an Environment Canada researcher, speaking anonymously for fear of reprisal, called it "a crude, heavy-handed approach to muzzle Canadian scientists."

This systemic muzzling has attracted censure from a broad swath of the world's scientific community. An editorial in the journal *Nature* insisted it was "time to set scientists free," and numerous national and international science organizations attending an American Association for the Advancement of Science (AAAS) meeting in Vancouver in early 2012 sent a joint letter to the prime minister demanding "unfettered access" to government scientists and their work. The Canadian Science Writers' Association (CWSA) was among the signatories, and its officers have been outspoken in the Canadian media about their disdain for the new policy. "The purpose of our taxes is not to prevent government scientists from talking to a free press and explaining to journalists what they've found in a way that they make sure doesn't upset government priorities and programs," CWSA president Stephen Strauss told me a few months after the meeting. "That's not the purpose of us funding this research. We are funding the research to find out how nature works, how things work. And because they are our servants—that is, the scientists—they should be able to speak freely to the press and explain what they have found with our money."

ALL THIS DISCORD was mere prelude to Bill C-38, the omnibus budget bill that became law in June 2012. C-38 came cloaked in the stentorian language of budget cuts and program cancellations, but its revolutionary intent is evident in the fine details scattered throughout its 400-plus pages. The bill is best understood as a categorical dismissal of the primacy of basic science in government affairs and a declaration of open hostility to fact-finding in the environmental sciences. With C-38, a new era in Canadian politics began.

To be sure, there had been other glib dismissals of evidence and reason in the preceding years. In 2010, for example, the minority Conservative government brought in an AIDS research policy that provided scant funding for treatment programs and chose not to sign the UN's Vienna Declaration, which called for "evidence-based approaches" to drug policy aimed at reducing the spread of the disease. Dr. Julio Montaner, director of the B.C. Centre for Excellence in HIV/AIDS, described the federal government's AIDS policy as "criminal neglect." The government's sweeping crime bill, passed the same year, overlooked all available evidence about current Canadian crime rates, dismissed the concerns of numerous law-enforcement experts and professional organizations, and ignored evidence of more effective approaches to crime reduction. The bill's loose relationship with facts had even given rise to the singular spectacle of Treasury Board Minister Stockwell Day claiming that there had been a shocking increase in "unreported crime."

Bill C-38, though, was a much more ambitious beast than these minority-era policies. It was a unilateral dismissal of more than a century of environmental policy built on the best available evidence. It was the first frontal assault in the government's war on science itself.

Not that science was the sole target of C-38. Under the guise of careful spending in rough economic times, the government's omnibus budget bill extended its reach far beyond basic research. It slashed funding for a broad range of Aboriginal initiatives, cutting healthcare budgets for numerous First Nations groups and eliminating its funding for the First Nations Governance Institute and the First Nations Statistical Institute. The bill reduced Parks Canada's budget so significantly that during the winter of 2013 many national parks had to close their visitor centres and rely on volunteers to clear snow at access points and on recreational trails. The Canadian Food Inspection Agency, Agriculture Canada, and the Department of Foreign Affairs and International Trade all endured significant staff reductions. But the most vicious cuts were aimed at science—in particular, those programs and departments that fund the basic research, data-gathering, field-monitoring, and communications tools that previous governments had assembled over the years to provide sound environmental stewardship. And the bill's cuts accompanied a wide range of reductions and eliminations earlier in Harper's tenure, demonstrating a clear pattern of disdain for the government's stewardship role.

"This wasn't tinkering," federal environment commissioner Scott Vaughan told Postmedia News, as he began to review the impact of C-38 in the months after its passage in his role as the government's internal environmental watchdog. "This was a wholesale game changer."

By the time Bill C-38 was passed, the Conservative government had already eliminated the position of National Science Adviser and begun to overhaul the National Research Council, restructuring departments and eliminating funding streams to better perform as a "concierge" service for business and industry. (This was secretary of state for Science and Technology Gary Goodyear's phrase. Goodyear had distinguished himself soon after his appointment by refusing to tell the *Globe and Mail* whether he acknowledged the scientific validity of evolution.) Changes in priorities and reductions in funding to the Natural Sciences and Engineering Research Council (NSERC), the primary federal funding body for basic science research at Canada's public universities, had also caused widespread dismay among scientists. By 2012 there were few government research programs and departments in any scientific field that hadn't been shaken by the Conservative government's aggressive agenda. But all this was merely an overture to the C-38 cuts.

Bill C-38 fundamentally rewrote the Canadian Fisheries Act, reducing its mandate from all fish habitat to only that of "valuable" fish populations, leaving more than half of Canada's freshwater fish species and 80 percent of species at risk of extinction unprotected. The omnibus budget also repealed the Canadian Environmental Assessment Act and amended the Species at Risk Act and the Navigable Waters Protection Act, the latter of which was further revised a few months later in Bill C-45, 2012's second omnibus budget bill. (Before C-45, the Navigable Waters Protection Act provided environmental oversight for nearly 3 million bodies of water across Canada; after the bill became law, just 162 remained protected.)

The list of C-38's cuts and closures went on and on. In addition to shutting down the Experimental Lakes Area, the bill axed the National Roundtable on the Environment and the Economy (NRTEE) and eviscerated the Canadian Foundation for Climate and Atmospheric Sciences (CFCAS). The cuts to CFCAS necessitated the closure of the Polar Environment Atmospheric Research Laboratory (PEARL), Canada's only research and data-gathering facility in the High Arctic. The bill's myriad hacks and slices shut down oil-spill response stations in northern British Columbia, slashed staff at Department of Fisheries monitoring stations nationwide, and triggered the summary abandonment of 492 environmental impact assessments on a broad range of proposed industrial projects across the country. One of the bill's few new investments on the environmental front was an $8-million line item to permit Revenue Canada to perform a higher volume of audits on environmental NGOs, ostensibly to uncover what Natural Resources minister Joe Oliver claimed was widespread overspending on political activities in violation of their charitable status. (By the end of the first year under this new audit program, Revenue Canada had spent $5 million on almost nine hundred NGO audits to uncover a single organization in violation of the spending rule for political activities—Physicians for Global Survival, a nuclear disarmament group.)

Critics of Bill C-38 responded with widespread protest and censure. An open letter to the government signed by four former Fisheries ministers (two of them Progressive Conservatives) urged the government to reconsider its overhaul of the Fisheries Act. Scientists across Canada wrote letters and op-eds in protest at the bill's multitude of cuts to scientific research, environmental monitoring, and First Nations programs. And in Parliament, Green Party leader Elizabeth May tabled more than three hundred amendments to the bill, forcing a marathon twenty-two-hour voting session. The result, though, was never in doubt. This was not a government that reconsidered its decisions, especially not those shoving science aside in the name of achieving its political goals. "The kindest thing I can say," the University of Alberta's David Schindler told Postmedia News, "is that these people don't know enough about science to know the value of what they are cutting."

The many cuts and cancellations contained in Bill C-38 were not determined by a thorough review of the work they funded or the value it might provide (or fail to provide) to the Canadian public. The bulk of the bill's content was the work of the "Group of Nine"—a handful of Cabinet ministers and other Conservative loyalists who met in private a couple of nights a week in the last three months of 2011. Prior to their political careers, the co-chairs of the group had been a lawyer and a farmer, and the other members included three lawyers, an investment banker, a fighter pilot, a management consultant and school administrator, and a long-time Conservative party staffer. (To be fair, one of the lawyers present—Labour minister Lisa Raitt—holds a master's degree in environmental biochemical toxicology.)

"This kind of review is not just about finding savings per se, although that is important," Tony Clement told a partisan crowd at the Manning Centre for Democracy in Calgary shortly before the bill was passed. "It is also a tool by which we can help to modernize government. . . . We have to ingrain this idea of efficient and constrained use of tax dollars on a day-to-day basis at every level—from the politician all the way down to the proverbial mail clerk, to every level of bureaucracy." The Group of Nine consulted no mail clerks or bureaucrats—or scientists, for that matter—as they conducted their review. And whatever the process, the apparent conclusion was that to modernize government, it was necessary to curtail its recourse to scientific evidence.

Throughout Stephen Harper's rule there has been endless speculation about his "hidden agenda." But his most radical agenda has been hidden in plain sight, and Bill C-38 provided its most transparent expression. The bill's numerous budget cuts and program closures, combined with other reductions and dismissals since Harper first took the prime minister's oath in 2006, carve a clear pattern on the body of Canadian government.

"The idea is simple and straightforward: to make Canada the most attractive country in the world for resource investment and development, and to enhance our world-class protection of the environment today for future generations of Canadians." This was how Christopher Plunkett, the government's spokesperson in the U.S., explained the aim of Bill C-38 to the *Washington Post*. The order of the two priorities is telling, particularly after the tabling of a budget containing cuts to nearly every aspect of the environmental protection regime Plunkett bragged about, cuts that were opposed by everyone from former Fisheries ministers to scores of the scientists and other civil servants whose sole occupation was to administer and enhance Canada's federal stewardship regime. The first clause in Plunkett's statement—the one about accelerating resource development—is the actual policy goal. The second clause, talking about future generations, feigns allegiance to the tradition of environmental stewardship on which Canada has long prided itself. Stephen Harper's agenda is hidden, if at all, only in such rhetorical nods to priorities it no longer honours.

So what is the Harper agenda? Above all else, Harper's regime has consistently aimed to diminish the government's role in environmental stewardship in three simple ways: (1) by reducing the capacity of the government to gather basic data about the status and health of the Canadian environment, particularly in places where lucrative resource development is expected; (2) by shrinking or eliminating offices and organizations—both government agencies and NGOs—that monitor and analyze that evidence and respond to emergencies; and (3) by seizing control of the communications channels by which all of the above report their findings to the Canadian public. The ultimate goal is equally clear: to reduce the government's ability to see and respond to the impacts of its policies, especially those related to resource extraction. There's a sort of blunt, brazen genius in how poorly the agenda is hidden—the statement celebrating the new policy goal often points at the goal to be discarded, all but begging its audience to notice the discrepancies. Surely a government trying to disguise its intent would not be so obvious, so clumsy.

If you look at the impacts of Bill C-38 and its policy cohort on environmental science, the agenda's real aims become all too clear. From the freshwater lakes of the ELA to the High Arctic monitored by PEARL to the coastal waters where the Department of Fisheries and Oceans does its data-gathering to the creeks and rivers once protected by the Navigable Waters Act, the Harper agenda has steadily diminished the government's capacity to analyze and assess the health of the natural world. (The elimination of the long-form census has had a similar impact on the social sciences.) By eliminating the pesky reports produced by organizations like CFCAS and the NRTEE (which repeatedly warned the government of its failure to adequately address the challenge of climate change), by closing down offices like the ones occupied by the DFO's marine contaminants program and Environment Canada's Environmental Emergencies Program, and by demonizing environmental groups and freezing their funding while they undergo nuisance audits, the Harper agenda has dramatically reduced the nation's capacity to respond to crisis. And by muzzling scientists and obliging all government staff to direct queries to media officers armed with talking points, the Harper agenda has made it substantially more difficult for scientists to tell Canadians what is happening to the ecosystems in which they live and breathe. Do No Science, Hear No Science, Speak No Science—that is the Harper agenda. And if this agenda is most evident and most pronounced in environmental science, that is simply because it is the field most likely to uncover evidence that the government's paramount goal—to free the country's resource extraction industries from regulatory oversight in the name of rapid expansion—is wrongheaded, reckless, and damaging.

Speaking at Carleton University in Ottawa in September 2012, pollster Allan Gregg delivered a summary of the Harper agenda at work in Bill C-38. "This was no random act of downsizing," Gregg said, "but a deliberate attempt to obliterate certain activities that were previously viewed as a legitimate part of government decision-making—namely, using research, science, and evidence as the basis to make policy decisions. It also amounted to an attempt to eliminate anyone who might use science, facts, and evidence to challenge government policies."

Gregg also noted that the repercussions of this transformation reverberated far beyond any given fiscal year or election cycle. "More than anything else, societal progress has been advanced by enlightened public policy that marshals our collective resources towards a larger public good. Once again it has been reason and scientific evidence that have delineated effective from ineffective policy. We have discovered that effective solutions can only be generated when they correspond to an accurate understanding of the problems they are designed to solve. Evidence, facts, and reason therefore form the *sine qua non* of not only good policy but good government... It seems as though our government's use of evidence and facts as the bases of policy is declining, and in their place, dogma, whim, and political expediency are on the rise."

Afterward, Gregg posted the text of the speech to his blog. He gave it a provocative, Orwellian title: "1984 in 2012." The text quickly found a wide audience on the Internet. In the wake of C-38 and the Death of Evidence march just a few weeks before Gregg's speech, Canadians finally appeared to be seeing the Harper agenda clearly. It was a bureaucratic war on science, on reason—on the very foundations of Enlightenment thought. As Gregg noted in a follow-up essay in the *Toronto Star*, the Conservative government's betrayal of the primacy of science shifted the parameters of political debate in Canada from right versus left and markets versus regulators to the very beginnings of the whole democratic project in the eighteenth century, when "Enlightenment thinkers began extolling the importance of reason as the linchpin of liberty and progress." More than two centuries later, the Harper agenda has resurrected the conflict between objective reason and royal prerogative as final arbiter of the greater good.

There have always been heated debates about policy and its goals in Canadian politics. Enlightened debate, after all, is the feedstock of democracy. But C-38, the fact-deficient crime bill, the discarded long-form census—these have been radical steps outside those accepted parameters. There's a famous quote, usually attributed to former U.S. senator Daniel Patrick Moynihan, that goes "You're entitled to your own opinions, but you're not entitled to your own facts." This truth is understood in every high school debating session, in every master's thesis and every government report. And it was understood in Canada's Parliament—until the arrival of the Harper agenda.

It is telling that one of the most damning screeds against that agenda comes from a pollster who built his career in the Progressive Conservative Party, that the most prominent opposition to C-38 came from former Progressive Conservative Fisheries ministers, and that the most striking piece of political theatre since C-38 passed was a march by lab-coated scientists. Protests on Parliament Hill are as much a part of Ottawa's landscape as the tulip blooms each spring, and deposed Cabinet ministers have often pointed out the perceived failures of their successors. And, of course, pollsters whose star candidates (Kim Campbell in Gregg's case) lose hard sometimes turn bitterly against the ones who pick up the pieces and make something better. But this current wave of criticism was not partisan. None of it had anything to do with politics at all—at least, not politics as understood by those outside the PMO and the Group of Nine. This wasn't an argument about who held sway in the House of Commons but rather an affirmation of the sanctity of the foundations on which the whole House is built. These were defences of basic principles—of science, of reason, of good government and the enlightened policy on which it stands. These were the alarm calls of insiders who had suddenly seen a warning gauge leap into the red zone, a place they had never before seen it go.

THE WILFUL BLINDNESS imposed on Canadian government by Bill C-38 will have profound and lasting implications for the gathering of data and the formulation of evidence-based policy. Consider a single item in the bill's extensive list of cuts: the decision not to renew funding for CFCAS. Although CFCAS still technically exists as an organization (under a new name, the Canadian Climate Forum), it can no longer perform its most important function—to distribute grants to researchers studying climate science. One of its most significant grant recipients was the research team working at PEARL, Canada's only research station in the High Arctic.

PEARL is a modest facility, a simple red cube deposited on the desolate permafrost of Ellesmere Island. It sits well north of the Arctic Circle, just north of 80 degrees latitude and a mere 1,100 kilometres south of the North Pole. It contains four research laboratories, and its roof is studded with instruments for measuring atmospheric conditions. It was opened in 1992 and has operated year-round since 2005. PEARL provided the only Canadian window on weather and climate during that yawning four-month interval of unbroken winter darkness known as "the Arctic night." The world in which PEARL operates is so fragile and still, and the impact of the slightest change so profound, that the tire tracks of the construction equipment used to build the facility are still visible on the ground around it.

In exchange for this one-of-a-kind research facility, which produced dozens of globally important scientific papers and trained more than fifty scientists in the atmospheric sciences in its first six years of year-round operation, PEARL required $1.5 million per year in funding from CFCAS. After Bill C-38 passed, PEARL's money vanished and its directors had to scramble for new sources of funding simply to keep the lab open during the warmer months. The lab's small light in the Arctic night appeared to have gone out indefinitely.

Despite the mounting threat to funding for environmental science in the Harper years, the impending closure still took PEARL's principal investigator, James Drummond of Dalhousie University, by surprise. Wasn't this the government that had made "Arctic sovereignty" a priority? Wouldn't information on atmospheric chemistry, climate, and the ozone layer in the Arctic be useful in pursuing that goal? "Sovereignty is not just a matter of military presence," Drummond told me. "Sovereignty is a matter of knowing the region, of understanding the region, and being able to effectively represent that region to the rest of the world. And we're falling down on the job." The indifference toward PEARL's work struck Drummond as particularly odd, given the government's stated interest in oil exploration in the Arctic. "If we are going to do resource extraction in the Arctic, which seems to be something that we're told is going to happen, then we really ought to know what's going on in detail, because the Arctic is an extremely fragile region."

The issue with PEARL was not financial. After all, the government had no trouble finding $270,000 to fund a single six-week mission to find the remains of John Franklin's doomed Arctic voyage and gather extensive sonar data on the Arctic Ocean's sea floor. And just months after slashing CFCAS's funding, Prime Minister Harper travelled to Cambridge Bay, 1,800 kilometres south of PEARL, to announce $200 million in funding for a new Canadian High Arctic Research Station (CHARS), which is expected to begin operations by 2017. "The right place to do research about the North is in the North," Harper told the CBC in Cambridge Bay. "The Canadian High

Arctic Research Station will also enhance Canada's visible presence in the Arctic. In this way science and sovereignty are entwined." On the official CHARS website, the first priority listed is "Resource Development." The third priority—after "Exercising Sovereignty"—is "Environmental Stewardship & Climate Change."

James Drummond seemed baffled that a government that had put such a high priority on Arctic oil exploration would shutter a facility providing the only window on the region and invest so much money in a new installation hundreds of kilometres south of the Arctic's richest oil and gas deposits. His confusion was understandable; he was thinking like a scientist, hunting for logic and reason of a sort no longer welcome in Stephen Harper's government. There is, however, a logic at work in the closure of PEARL and the lavish funding for CHARS and the Franklin ghost-chasing mission. Here it is: if you intend to begin large-scale exploration for resources in the High Arctic but don't want to know whether it's safe and ecologically sound to exploit those resources, the Harper agenda makes perfect sense. And the only logical explanation for such reckless spending on a quixotic hunt for John Franklin's remains is that the primary goal of the project must be not to find a 150-year-old corpse but to gather data to support CHARS's top priority—resource development. Again, seen through the lens of the Harper agenda, the inexplicable seems almost obvious. Esoteric lab work on Ellesmere Island can only tell you that exploration in the High Arctic is extremely dangerous, that the potential for catastrophic oil spills is greater than anywhere else on earth, and that the final result of exploiting the resource would be to intensify a climate change catastrophe that is already well underway in the Far North. But innocuous sonar data and a state-of-the-art facility too far south to offer any clue about the climate in the vicinity of the Arctic's biggest oil deposits—these are the tools of a government with every intention of rushing blind into the frozen abyss.

In May 2013—with little advance notice or explanation—the federal government restored funding for PEARL. The facility's future had hung in limbo for more than a year. It had been operating only part-time, creating damaging gaps in its data record. And then, with the arbitrary timing of a feudal lord's emissary, the government's secretary of state for Science and Technology, Gary Goodyear, announced that five years of operational money had been found under a new program created to take on some of CFCAS's old roles. The only possible clue to the government's renewed interest in PEARL was that the announcement came the day after Stephen Harper delivered a speech to the Council on Foreign Relations in New York, in which he attempted to convince influential members of the policy establishment in the U.S. that his government was a dedicated and responsible environmental steward. The Obama Administration was debating whether to approve the construction of the Keystone XL oil pipeline at the time. PEARL evidently possessed little value to the Harper agenda in and of itself, but it was useful as window dressing on a public relations campaign.

There is an absurd aspect to the Harper agenda, something almost self-parodic about its blatant internal contradictions. It is almost as though there is a deliberate effort to throw up such a bizarre and voluminous array of outrages and inanities that they provide cover for the government's real work. In August 2011, for example, the Department of National Defence (DND) announced plans to begin developing a "stealth snowmobile" to enable covert military operations in the High Arctic. A princely $550,000—more than one-third of PEARL's annual operating budget—was being invested in the design of a prototype. Just weeks later, the Group of Nine would begin its cost-cutting work in Ottawa—which, as Tony Clement explained, intended to "ingrain this idea of efficient and constrained use of tax dollars on a day-to-day basis at every level—from the politician all the way down to the proverbial mail clerk, to every level of bureaucracy." Except DND, evidently.

Even the Harper agenda's absurdities, though, speak to a deeper crisis in good governance and the role of science in guiding it. Here's a case in point: in March 2012, the *Ottawa Citizen*'s Tom Spears sent a routine interview request to the National Research Council (NRC). Spears had heard about a joint research project in which the NRC and NASA were working together to investigate that quintessentially Canadian topic of snow. Specifically, the research team was studying what causes snow to fall in the quantities and densities it does. It turns out that while at the macro scale we know quite a lot about snowfall, the micro-scale variations are still fogged in mystery. Spears was hoping to interview a few scientists working on the project to gather a little more detail. It was a straightforward scientific-slice-of-life story, revealing some of the trench work carried out mostly without fanfare by the 23,000 scientists on the federal government's payroll. When Spears's request arrived at the NRC, the organization's own media officials characterized it as a "positive/informative" story.

The NASA team got back to Spears right away, and in the space of a single short phone call, he'd learned all he needed to know about their role in the project. Still, the story begged for a Canadian angle, so Spears waited for an NRC scientist to fill out the story. His email query to the NRC, however, spent nearly a full day pinging from inbox to inbox, as communications officers fretted over the wording of their background materials, "massaged" replies, and debated whether an interview was necessary. The most senior communications bureaucrat in an eleven-message email chain eventually decided it wasn't. The *Citizen* ran its story without comment from the NRC or anyone else in the Canadian government, making only passing mention of the NRC's participation.

Months later, in response to an Access to Information request filed by the *Citizen*, Spears received all fifty-two pages of byzantine bureaucratic brow-furrowing over the NRC's participation in the story. Spears had called before 10 AM the day before the story ran. The NRC's media team prepared a standard "request" email, which bounced between inboxes for several hours. At some point in the early afternoon, it was decided that an interview wasn't warranted, and then a chunk of text so bland it could have been a corporate brochure about the project went to seven inboxes to ensure that the language was permissible. Sometime after 4 PM, after Spears had filed his story, the NRC's marketing manager noticed that the briefing had neglected to include the fact that the Canadian portion of the project had been funded by the Canadian Space Agency. At this point the messages took on a tone of heightened urgency. Not only had they failed to take advantage of a rare opportunity to make Canadian scientists look good in a "positive/informative" story, they hadn't even managed to properly credit their partners.

There's no missing the absurdity in this "blizzard of bureaucracy," as the *Citizen* headline put it. The government of Canada—the Great White North—had been unwilling to discuss what it knew about snow. The subject of the study was no more controversial than a weather report, and Spears's request promised infinitely less friction than a routine Parliament Hill press scrum. So why the handwringing? Why the canned text? Why use "no comment" as a default setting? Again, the whole incident only makes sense if it is viewed through the distorted lens of the Harper agenda's logic: *science itself* is a problem in Stephen Harper's Ottawa. Like PR flacks for an oil company prone to spills, the government's media relations officers no longer trust scientists to avoid controversy, because the government's whole agenda rests on a baseline disdain for certain kinds of science—environmental science, especially. Snow is a weather phenomenon, an element of climate; discussing snowfall patterns just might lead to discussion of how those patterns are changing, how certain kinds of industrial activity are contributing to those changes. The entire field of environmental science is suspect. Best to say nothing at all.

"What you see is the kind of process that occurs when you've turned the information process into something that has to go through a bureaucratic grind," veteran science reporter Stephen Strauss told me. "You've heard about the Satanic mills of the Dickensian era? You're looking at the Satanic mills of Stephen Harper's information era."

The results might be laughable, but they are not trivial. The bureaucratic blizzard that paralyzed the NRC's media department is a grave distress signal from a government in which distrust of basic facts has thoroughly saturated its day-to-day operations. Even the most mundane work of government—or at least those branches that address the conditions of the natural world—is now subject to a top-down, command-and-control agenda in which the truth of a piece of information carries substantially less weight than its political ramifications. Inside the Ottawa information bubble, as in a snow globe, the snow falls only by the grace of the prime minister's guiding hand. Without it, no one feels safe to tell you how the weather is.

THE HARPER AGENDA'S impact on the basic operations of Canadian government should not be underestimated. It goes far beyond overzealous message discipline. It has created a government that will not listen to inconvenient facts or insubordinate experts on matters of vital policy. It has hobbled the ability of some critical departments to fulfill their mandates and do their jobs. It has blinded Canadians to who we are, where we live, what our government is doing, and why it matters.

Consider some of the specifics. When the Conservative government stopped funding basic health care for refugee claimants, eight major national healthcare professional groups—everyone from the Canadian Medical Association to the Canadian Dental Association, collectively representing virtually the entirety of Canada's healthcare system—signed a letter urging them to reinstate the funding. The request from these experts was ignored with such contempt that individual doctors began showing up at government press conferences in their lab coats to speak out—a phenomenon as unprecedented as the march of the lab coats on Parliament Hill. When the Joint Review Panel assessing Enbridge's proposal to build the Northern Gateway oil pipeline across northern British Columbia asked officials at the Department of Fisheries and Oceans (DFO) to forward its environmental impact assessments, the

DFO replied that it hadn't completed them yet and lacked the resources to prepare them before the panel's ruling. The panel would thus be obliged to rule half-blind on one of the most contentious infrastructure projects in a generation. Stephen Harper asserted that "science" would determine the pipeline ruling, but as retired DFO scientist Otto Langer told the CBC, "He's basically disembowelled the science."

The list goes on and on, forming a pointillist portrait of a government struck blind by its own hand. And the portrait reveals, to a keen eye, the motive behind the Conservative government's self-inflicted wounds.

Let's imagine the scene in Centre Block on those darkening nights in late fall 2011 when the Group of Nine met to give more than four hundred pages of shape and purpose to the omnibus budget bill. The whole vast apparatus of Canadian government had been laid out on the table, so Tony Clement has assured us, the assembled team tasked with finding $7 billion in unnecessary spending and superfluous programs to trim away. Agricultural marketing programs, consular offices, federal research labs tasked exclusively with developing new upgrading technology for the oil sands. Stealth snowmobiles and the ballooning F-35 contract. More than a quarter of a million dollars to find the remains of the Franklin Expedition, $16 million to blitz the country with "Canada's Economic Action Plan" advertising, more than $28 million to commemorate the bicentennial of the War of 1812. The management of vast parklands, billions in real estate holdings, billions more in highway construction and maintenance. All the stuff of a modern government in a prosperous industrial nation. By what forensic accounting technique does the Group of Nine's budgetary scalpel locate, for example, the continuation of a

$2-million-per-year research program coming up on its 45th anniversary? How was it that the demise of the Experimental Lakes Area became a priority of the Harper agenda?

Well, let's be clear about what the ELA does. It is one of the world's foremost freshwater research programs. In 1969, the year after it was founded, Lake Erie and many other freshwater ecosystems were choking on a reeking green scum of algae, and research at the ELA soon determined that phosphorus from industrial and municipal waste streams was the cause. As a result, governments around the world introduced regulations to reduce or eliminate phosphates in laundry detergents and in many other consumer and industrial products. Some years later, an ELA project determined unequivocally that acid rain was caused by air pollution, especially the toxic smoke produced by coal-fired power plants. As a direct result of those findings, Prime Minister Brian Mulroney and President George H.W. Bush met in 1991 to sign the comprehensive "acid rain treaty" in the very same Centre Block building where the Group of Nine would later gather.

In the time-honoured tradition of scientific research and enlightened government policy, the ELA was a federally funded program that discovered proof of the devastating toll that industrial processes were taking on freshwater ecosystems and provided the insight needed to create new environmental regulations to protect those ecosystems. In recent years, David Schindler of the University of Alberta—one of the ELA's co-founders and the architect of its trailblazing phosphate study—had become Canada's most prominent investigator of the ecological costs of oil extraction and processing in northern Alberta. Less than a year after Bill C-38 became law, researchers at Queen's University and Environment Canada, building on Schindler's research, published a report showing conclusively that carcinogenic runoff from bitumen mining was finding its way into freshwater ecosystems downstream from oil sands projects.

In the Enlightenment logic of governments past, this new study would oblige the government to pass new regulations for the oil industry. Such logic would also conclude that the ELA's work has never been more vital to the health of both the Canadian environment and the Canadian public. The logic of the Harper agenda, however, determined that the best response was to cancel the ELA's funding—all at once, as if in an emergency. The very same budget allocated $8 million in *new* funding to Revenue Canada, specifically so that it could perform more frequent and more robust audits of environmental groups of the sort that had been championing Schindler's findings and the ELA's work generally, even when the government chose not to do so.

There is, again, nothing hidden in this agenda. By its logic, it is worth spending four times as much taxpayer money to harass public interest groups as it would have cost to maintain a program that has demonstrated repeatedly over the past forty-four years that its research protects public health and the health of the Canadian environment. Silencing the government's sometimes vocal critics is worth $8 million; protecting Canada's irreplaceable freshwater resources isn't worth even a quarter of that. The government has made resource development its absolute top priority, and Canada's growth as an "energy superpower" is first among equals in the pursuit of that goal. In the absence of any real transparency, there are any number of possible reasons that the ELA and other basic environmental research and monitoring programs were hacked away by the Group of Nine. The most obvious reason, though, is that the government has an agenda, only partially stated. To accelerate the exploitation of Canada's resource wealth—that's the explicit part. To eliminate its ability to see the cost of this policy—that's the implicit part.

John Smol of Queen's University, the lead researcher on the study that linked bitumen mining with freshwater pollution, called its findings "the smoking gun"—indisputable proof that Alberta's oil industry had created an unacceptable risk to public and ecological health, that more stringent environmental controls were needed. It was a textbook case of enlightened policy development—the federal government, having funded data-gathering and expert analysis of a vital industrial sector, had discovered a shortcoming in its stewardship regime. Previous governments, whatever their political leanings, have interpreted such discoveries in a mostly uniform way. Environmental regulations for the oil sands would have to be rewritten, strengthened and clarified so that the industry could carry on with its growth within parameters that properly safeguarded the public good.

The Harper agenda, however, interprets a smoking gun differently. The problem is not the discovery of telling evidence but the very fact that the issue was investigated at all. The agenda is better served if there is no new information—no field samples, no analysis, no conclusion, nothing reported in the press or interpreted by experts. Only by destroying all the firearms, in other words, can you guarantee there will be no more smoking guns.

3

FROM DAWN TO DUSK

The Scientific Tradition
in Canadian Government
1603–2011

S INCE THE earliest days of European contact, two
broad scientific traditions—let's call them the
mercantile tradition and the laboratory tradition—have co-existed in a loose and mostly peaceable tension
as Canada progressed from colony to dominion to modern nation. As a vast, sparsely populated, and resource-rich
land mass, Canada has long inspired a primarily mercantile approach to science, practical and short-term in worldview and single-minded in its obsession with the exploitation
of the country's natural resources. This is the first tradition,
the motivation for colonization and the origin of Canadian
nationhood, and it has always been the dominant one.

At the same time, Canada was a New World entity, a place
far from the biases, traditions, and superstitions of the feudal European order and heavily influenced by the explosive

experiments in Enlightenment thought occurring in the dynamic democracy to its south. Although later to emerge, this enlightened tradition—the tradition of the research lab, often driven by nothing more than curiosity and the human impulse to explore, enabled by democracy's abundant freedoms—is the post-colonial nation's stronger self, the ideal for which it has long strived. Canada is both a mercantile warehouse and an Enlightenment laboratory, and its scientific tradition has been defined by the balance between the two approaches.

The dual role was in place from the moment Samuel de Champlain first set foot on North American soil in 1603. Champlain was a navigator, cartographer, and explorer—essentially scientific trades by the standards of the early seventeenth century—who had hitched a ride on the fur-trading expedition of François Gravé, Sieur du Pont. Unlike Sieur du Pont, Champlain came not just to harvest natural resources but to establish a colony. He was at the service of the French crown, but his gaze looked beyond his homeland's treasury to the western frontier. He founded Quebec, the first permanent settlement in New France, in 1608, and settled there in 1620. His ethnographic explorations of the new colony eventually took him as far west as Lake Huron. He died in Quebec City in 1635, the same year the New World's first institution of higher learning, the Collège des Jésuites, was established there. (Harvard College welcomed its first students in the Massachusetts colony the following year.)

Although fur-trading, fishing and other commercial pursuits for the benefit of distant thrones were to dominate the public life of the Canadian colonies from Champlain's day until well into the 1800s (if not the 1900s), the Enlightenment pursuit of knowledge and science remained a strong if intermittent part of the mix. Many of Canada's first scientists were Jesuit missionaries, educated at the Collège de La Flèche in France, which had also taught Descartes. Although their motives and goals were mostly spiritual, the Jesuits were skilled astronomers and natural historians who kept meticulous records of Canadian geography and geology and catalogued its abundant flora and fauna instead of merely harvesting them.

For one shining moment under the brief governorship of Roland-Michel Barrin de La Galissonière, from 1747 to 1749, New France even found itself at the Enlightenment's vanguard. La Galissonière was a keen supporter of Enlightenment ideals and a peer of the Académie royale des sciences in Paris. He ordered his military officers to supply him with a steady flow of new discoveries, feeding dreams of the birth of Sir Francis Bacon's Utopian "New Atlantis" in the Canadian wilderness. Alas, La Galissonière was summarily recalled to France in just the third year of his reign, and later colonial governors didn't share his passion for pure knowledge.

The British rulers of Canada after the 1763 conquest were even less interested in science for its own sake. They viewed the Canadian wilderness as a colossal warehouse, and they were awed mainly by its extraordinary abundance, its inexhaustible supplies of treasure to plunder and load into the cargo holds of Europe-bound ships. Beaver pelts and cod, salmon and logs, wheat and gold and nickel—these were for many generations the chief obsessions of Canada's applied scientists, its engineers and technicians, and especially its trappers and loggers and farmers and miners and drillers. When the twenty-first-century government of Stephen Harper makes its common-sense pitch to Canadians to shrug off cumbersome environmental regulations for the sake of booming enterprise, to harvest the lucrative bounty today and leave the thorny problem of stewardship for some later colonial master, the argument resonates with Canada's oldest political tradition: the counting-house science of exploitation. Harper's Canada is a freelance colonial outpost, its patron the globalized marketplace, its industrial tools and harvesting expertise available to the highest bidder (or the one least bothered by its contempt for twenty-first-century notions of stewardship).

In his seminal 1930 study of Canadian economic history, *The Fur Trade in Canada*, Harold Innis famously lamented the colonial legacy carved across the landscape through the relentless extraction and export of natural resources for short-term gain. Innis argued that it relegated the country to the marginal status of "hewers of wood and drawers of water" in the wider European and North American economies. And to be sure, this was the predominant tradition in Canadian commerce and Canadian science well into the twentieth century. But the Enlightenment ideals and ambitions of La Galissonière never vanished entirely from the scene, and as Canada became a more democratic nation and its population swelled with immigrants in the nineteenth century, the idea of science as the objective foundation of good government gained credence. In particular, the arrival of growing numbers of Scottish immigrants, starting in the 1820s, changed Canadian science forever.

Scotland was the centre of British science in the nineteenth century, its great universities at Edinburgh and Glasgow considered superior even to Oxford and Cambridge in medicine, engineering, and the natural sciences. In the wake of the immigrant wave that brought scores of educated Scots to Canada, scientific societies soon sprang up in cities throughout the Canadian colonies, and the influx of Scottish professors had a profound effect on the teaching at Canadian universities. In 1842, Scottish-born geologist William Edmond Logan founded the Geological Survey of Canada; in 1849, Scottish-born engineer Sandford Fleming co-founded the Royal Canadian Institute with Irish architect Kivas Tully. Together, these two institutions became the pillars of Confederation-era science in Canada—alongside the anglophile Royal Society of Canada—and blazed the trail for the fertile era of government-funded research to follow in the twentieth century.

As early as the 1850s, Canada's scientific achievements—including Logan's singular collection of Canadian rock and mineral samples, which were celebrated at international exhibitions in London and Paris—had become a source of national pride. The second half of the nineteenth century would soon see many other landmark moments for Canada's fledgling scientific community. When the British government abandoned its magnetic observatory in Toronto in 1853, the Royal Canadian Institute took it over and carried on with the data-gathering. The Institute would also play host to the first public presentation of Sandford Fleming's concept of standard time in 1879, and its pioneering conservation efforts led to the establishment of Algonquin Park (the Dominion's first provincial park) in 1893. Canada's Royal Society formed in 1882, and in Montreal in 1884 it hosted the first meeting of the British Association for the Advancement of Science outside the United Kingdom. The Central Experimental Farm, whose research would yield the legendary Marquis Wheat strain and help turn Canada into one of the world's most abundant breadbaskets, was established by the federal government in Ottawa in 1886. The country's first marine biological research station opened in 1899. Ernest Rutherford began the work that earned him a Nobel Prize and invented the field of nuclear physics at McGill University in 1900. And the Geological Survey's hunt for coal placed Canada at the front ranks of the young field of paleontology.

If most Canadians entered the twentieth century as hewers of wood and drawers of water (and diggers of coal and loggers of timber and planters of wheat), the young country had also developed some remarkably egg-headed sidelights and high-minded hobbies. And in the progressive era just then dawning, Canada's knack for world-class science would find its first true champion in Parliament.

Sir Robert Borden was a solemn figure and a reluctant leader, a politician so lacking in charisma that on his retirement from public office a contemporary journalist remarked that his biography would "bore the people to death." When Sir Charles Tupper chose him to be his successor at the helm of the federal Conservative Party in 1896, Borden seemed destined to become a footnote to Canadian history. He described his appointment as "an absurdity for the party and madness for me." The signal achievement of his first decade and a half as leader was a series of epic defeats at the hand of Sir Wilfrid Laurier and the ascendant Liberals.

But if Borden lacked the dynamism of a Macdonald or a Laurier, he nevertheless harboured great ambitions. Borden was a committed progressive-era reformer, and as Conservative leader he aimed for nothing less than a wholesale transformation of Canada's public policy apparatus. He planned to eliminate the crony capitalism, corruption, and crass patronage that defined the country's nineteenth-century civil service and replace it with rational, efficient modern agencies, established at arm's length from the partisan Parliament and deploying "neutral competence" on matters of public policy.

(For Borden, *neutral competence* passed for a rallying cry.) Railroads and ports, public utilities and telecommunications— all should be managed by non-partisan bureaucrats who earned their positions based on their skills and rose through the government's ranks on their own merits, not through their political loyalties.

Civil service reform began under Laurier's Liberals, who were defending themselves against mounting charges of corruption, but Borden's election as prime minister in 1911 marked the real dawn of Canada's progressive era. Borden's government soon established arm's-length agencies to oversee the National Archives and the formulation of tariffs, and the reformist Conservatives likely would have gone further if not for the financial collapse of the railways and the outbreak of the First World War. During the war, Borden set up the Honorary Advisory Council for Scientific and Industrial Research, which would eventually morph into the National Research Council. The Civil Service Act of 1918, passed immediately after the war's end, ushered in the rest of Borden's ambitious reform agenda, establishing the ground rules under which Canada's civil service and government-funded science would operate for the rest of the century.

As Enlightenment thinkers from Kant onward had recommended, the light of reason and the revelations of science would form the foundations of public policy, implemented by a law-making body well informed by the best scientific expertise and objective data it could obtain. The partisans would continue to duke it out in the House of Commons, of course, but when it came to writing laws, managing departments, and conducting research in the public interest, reason and evidence would trump ideological arguments and short-term political goals. Although the system was never flawless, its basic assumptions informed public policy in Canada more often than not—until the election of its twenty-second prime minister, Stephen Harper, in 2006.

THE ELA IS both a monument to the wisdom of the progressive tradition established by Borden and a case study in its legacy—its successes and limits, the enduring tension between mercantile and laboratory traditions, and the eclipse of the latter by the former in the Harper era. The fifty-eight small lakes, carved onto the face of the Canadian Shield in northern Ontario amid thousands of others by the retreat of the glaciers at the end of the last ice age, were first set aside for research in 1968. The founding director was a brilliant, ambitious Rhodes Scholar named David Schindler, and the new research facility was placed under the auspices of the Fisheries Research Board, a grand old progressive institution in Borden's arm's-length tradition that had grown out of the management of Canada's first turn-of-the-century marine research stations.

The ELA was a research laboratory on a scale without precedent in the annals of aquatic science in Canada or anywhere else. Instead of test tubes and field samples, Schindler and his colleagues had the whole freshwater ecosystem in all its biodiverse richness to work with. The unique facility soon yielded singular scientific dividends. One of the great environmental horrors of the late 1960s and early 1970s was the eutrophication of North America's lakes—the explosive growth of enormous algal blooms fed by excess nutrients from industrial and agricultural runoff. Lake Erie, in particular, had become a byword for toxic pollution in those years, its surface thick with green slime, the rivers that fed it choked with the industrial runoff of America's industrial heartland. (The surface of the Cuyahoga River in Ohio, which flows into Lake Erie, was so thick with oily pollution that it caught fire in 1969, achieving global infamy.)

Scientists suspected that phosphorus was to blame for the eutrophication. Heavy industry and large-scale commercial agriculture both used vast quantities of phosphates, and every household in America added to the phosphate flow each time dishes or clothes were washed. In the absence of conclusive evidence, however, industry lobbyists held sway with their claims of innocence. In the ELA, David Schindler was given the unique opportunity to duplicate Lake Erie's inundation with phosphorus in a controlled setting. In 1973 he erected an underwater barrier across a narrow channel in Lake 226, creating two separate basins, like a massive double sink. Schindler added carbon and nitrogen (both present in elevated quantities in eutrophic lakes) to both sides of the barrier; on one side, he also added an enormous amount of phosphorus. Aerial photos of Lake 226 that summer told the shorthand tale that Schindler would detail at greater length in a landmark paper for the journal *Science* the following year. The water in the basin without phosphorus remained the usual deep indigo; the surface of the phosphorus-rich basin had gone pea-soup green in just months, as algae gorged themselves on their favourite industrial pollutant.

Schindler's study established with scientific certainty that phosphates in fertilizers and industrial and household detergents, which were washing into Lake Erie and countless other lakes in heavily populated regions the world over, were the cause of the algal blooms that were choking the life out of many of the planet's aquatic ecosystems. Within years, governments around the world began to phase out and ban phosphates from a range of industrial processes and household products. And Canada's living lab deep in the wilderness of northern Ontario established itself practically overnight as the most important freshwater research facility on earth.

Just two years after the publication of David Schindler's eutrophication paper, he and his ELA colleagues initiated a series of experiments at Lake 223 that would lead to their next world-shaking discovery—and serve as one of the crowning achievements of Canada's golden age of environmental stewardship. In 1976, ELA scientists inundated Lake 223 with sulphuric acid, imitating the effects of acid rain. Biodiversity in the lake declined steadily over the next few years; by 1983, lake trout and white suckers verged on extinction as a result of disruptions to their reproductive cycles. The following year, Schindler and his team reduced Lake 223's acidity, simulating a reduction in the air pollution suspected of causing acid rain. White suckers began reproducing again almost immediately. As a result, lake trout regained a key food source and their populations soon recovered as well.

The paper documenting the experiments at Lake 223 appeared in *Science* in 1985. In 1988, as Prime Minister Brian Mulroney repeatedly pressed the U.S. government to address the acid rain problem, Schindler testified at a range of legislative hearings in the United States and Canada and told the ELA's alarming acidification tale everywhere from the *New York Times* to NBC's *Today* show. In 1991 the International Joint Commission—the organization that manages issues in boundary waters between the U.S. and Canada—signed its landmark Air Quality Agreement, better known as the "acid rain treaty." Once again, the ELA had fundamentally reconfigured public policy on an international scale.

The acid rain treaty came at the end of probably the most celebrated and successful era in Canadian environmental stewardship, culminating in a string of global triumphs under the remarkably effective Environment ministry of the Mulroney years. Partly as a result of the work accomplished during this trailblazing age, Canadians still tend to see the mythic good Canadian reflected back at them when they peer into the dark, mirrored surface of the nation's countless lakes. The quiet, polite Canadian. The peacekeeper, the steadfast Mountie, the voyageur. The responsible park warden, the global leader on key ecological issues. They see an unbroken chain of smart, enlightened stewardship befitting a nation whose territory remains, as ever, more wilderness than civilization. If clean lakes and deep, wild woods and towering snowcapped mountains aren't Canadian, then what is? If we're not the good guys, who else possibly could be? Surely that all remains, timeless and immutable, despite the hue and cry of the protesters and summit-goers? This perception persists in large part because Canadians continue to bask in the reflected glow of a golden age in environmental science and environmental stewardship—an age now in eclipse.

As with the ELA itself, Canada's rise to the front ranks of the global environmental research community traces its origins to the early 1970s. During the 1960s, the nation's universities had experienced an enormous expansion, churning out PhDs in the sciences at an unprecedented volume. Amid the OPEC energy crises and budget crunches of the 1970s, however, research jobs became scarce on campus but relatively plentiful in government, and bright scientific minds flocked to government labs and bureaucratic payrolls as never before— especially in the young, burgeoning field of environmental science. "When I started in the 1970s, it was universities that were regarded as not having much excellence," David Schindler told me. "All of the excellence in aquatic research, at least, and air quality research, was thought to be with the federal government."

By the end of the decade, Canadian environmental scientists on the government's payroll routinely achieved major breakthroughs, and they set the international agenda for some of the planet's most pressing environmental issues throughout the 1980s. The very first initiatives to curb acid rain in North America were launched during Joe Clark's brief tenure as prime minister. A few years later, building on these efforts, Brian Mulroney's Progressive Conservative government established probably the single most effective environmental policy regime in Canadian history. Under Mulroney, the Canadian government brought in new legislation of its own to reduce sulphur dioxide emissions and began pressing the United States to take action on acid rain, culminating in the landmark 1991 treaty. Canadian legal and scientific experts took lead roles in the emerging global conversation about ozone depletion—sounding the alarm, establishing the intensity of the crisis, and serving as co-authors of a global agreement to take decisive action to reverse the growth of the hole in the ozone layer. The deal, signed by 46 countries in 1987 and since ratified by 151 more, was named for the place it was brokered, in homage to the country that had done the most to make it happen: the Montreal Protocol. Green Party leader Elizabeth May, a senior adviser in Mulroney's Environment department during those years, has called the agreement "the most significant global treaty to protect life on earth since the 1962 treaty to end atmospheric nuclear weapons testing."

The Canadian government's leadership on climate change under Brian Mulroney is, in retrospect, particularly impressive—both because the science was so new and complex and because the party Mulroney once led has become the global standard for sneering inaction on climate change since its merger with Stephen Harper's Canadian Alliance. There is no other issue on which the Canadian government's retreat from reason and evidence is more abundantly evident, no other critical policy question on which Canada has turned so fully away from its traditional allies and the scientific community it once led. A quarter-century before Stephen Harper's Conservatives reneged on Canada's commitment to the Kyoto Protocol, Brian Mulroney's Progressive Conservatives were among the central architects of the bureaucratic apparatus that produced it.

Rewind to the mid-1980s. In response to mounting evidence of the catastrophic ecological toll of industrial activity the world over, the UN established the World Commission on Environment and Development to respond to these planetary-scale problems. The commission—popularly known as the Brundtland Commission in honour of its forceful and effective chair, future Norwegian prime minister Gro Harlem Brundtland—produced a landmark 1987 report, "Our Common Future," which popularized the term "sustainable development" and initiated the international negotiating process that led to the 1992 Earth Summit in Rio de Janeiro and the Kyoto Protocol on greenhouse gas emissions.

Throughout, Canadian officials played vital roles. Key funding for the Brundtland Commission came from the Canadian government, and Canada's representative on the commission, Maurice Strong, served as secretary general of the Rio summit.

Even the Kyoto Protocol—later a deeply loathed *bête noire* for Stephen Harper's Conservatives—was modelled on the wildly successful Montreal Protocol negotiated by Brian Mulroney's Progressive Conservatives. And in the years in between, Mulroney and his parliamentary colleagues embarked on a series of high-level discussions aimed at tackling what has become the defining public policy issue of the twenty-first century.

The legacy of the Progressive Conservatives' Mulroney-era environmental policy work has been erased so completely from the organizational DNA of the contemporary Conservative Party of Canada that its actions on climate change in the late 1980s and early 1990s seem almost unbelievable, a dispatch from not just another era but another political universe. The global environmental science community's first collective response to the Brundtland Commission's 1987 report, for example, was at a 1988 conference in Toronto, co-sponsored by the government of Canada. The event's official title was "Our Changing Atmosphere: Implications for Global Security"; one participating scientist dubbed it "Woodstock for climatologists." Prime Minister Mulroney delivered the opening address, Brundtland spoke, and Canadian government scientists played the role of enthusiastic hosts. The conference was intended to cover not just climate change but all manner of atmospheric science issues, from ozone depletion to acid rain. So ferocious was Canada's reputation for environmental leadership at the time that U.S. officials were initially reluctant to attend, fearful their Canadian colleagues would use the conference to berate them for their failure to take action on the sulphur dioxide emissions causing acid rain (as Canada's own ELA scientists had established just a few years before).

The Toronto conference provided a vital bridge between the 1987 Brundtland report and the 1992 Earth Summit, and it closed with a consensus statement on climate change more forceful and unequivocal than anything that has passed the lips of a Harper-era Environment minister. "Humanity," it read, "is conducting an unintended, uncontrolled, globally pervasive experiment whose ultimate consequences are second only to global nuclear war." This was how scientists spoke at conferences featuring the keynotes of Conservative prime ministers in 1988.

The Intergovernmental Panel on Climate Change (IPCC), the world's premier organization for informing policy-makers about climate science, formed later the same year. And the second World Climate Conference, held in Geneva in 1990, again featured Canadian policy-makers and climate scientists in lead roles, free and keen to set the global pace for action on climate change. The Canadian government's delegation, more than a half-dozen members of Parliament in all, included not just Progressive Conservatives but also Liberals and New Democrats. Climate scientist John Stone, a senior Environment Canada staffer who later led the department's climate modelling group, recalled a spirit of collaborative teamwork all but impossible to imagine today, especially at a climate conference. "It was mixed, and it was very open. I remember sitting around the big table at our embassy in Geneva, where the meeting was held, and these members of Parliament from different parties quite openly discussing Canadian positions and things like this. I mean, nowadays at these framework conventions, they're very tightly controlled and nobody from other parties is invited."

From Toronto to Geneva to the Earth Summit in Rio, Canadian government scientists stood at the forefront of climate research, and the work itself sat high on the government's agenda and received generous support. "I had tens of millions of dollars of additional money to put into climate research," Stone told me. "And we were able to hire some really good people, and the climate research group that we had, the climate modelling group, was as good as any group around the world, most of whom were larger and better funded."

AS THE CLIMATE change conversation moved from research to consensus to policy action, Canadian leadership must have seemed inevitable. The government's ratification of the Kyoto Protocol under Jean Chrétien was a formality. Canadian environmental science was at an apex, well funded and internationally celebrated, riding the momentum of a series of important discoveries and policy successes. And yet the seeds of decline had already been planted. The evisceration had in some sense already begun. Once again, the ELA tells much of the story in microcosm, and the catastrophic collapse of Canada's cod fishery explains the rest.

For both the ELA and the fish, the troubles began with the consolidation of a variety of government departments under the broad banner of the newly formed Department of Fisheries and Oceans (DFO) in 1979. At its birth, the ELA had been administered by the Fisheries Research Board via the Freshwater Institute, an arm's-length research organization based in Winnipeg. Run and overseen almost exclusively by scientists in the proud laboratory tradition, the ELA stood outside near-term legislative decisions and mostly beyond the reach of Parliament, much as Robert Borden had envisioned it. And the Fisheries Research Board was itself an august and internationally respected scientific agency—ELA founder David Schindler remembers hearing about the board's stellar reputation as a graduate student in Europe in the early 1960s, when even the most cash-strapped universities made sure to keep their subscription to its journal current. It was taken for granted in those days that the world's leading water research came from Canada.

The DFO was a very different kind of organization. The Fisheries portfolio had bounced around through the 1970s, partnered first with Forestry, then with the Environment brief, until environmental science and monitoring was cleaved off into its own ministry and the whole great sweep of fisheries work—not just bureaucratic management of the industry but also aquatic science, the monitoring of fish stocks, and the management of research bodies such as the ELA—was brought under the banner of the DFO. The resulting department was big, bureaucratic, deeply beholden to parliamentary whim, and relentlessly commercial in its thinking. It had only a tangential interest in conservation and none at all in science for its own sake. Its purpose was to keep Canada's lucrative fishing industries—particularly the twin gems of cod and salmon—happy and rich, and to keep the MPs who oversaw the harvest in office. Its senior brass knew next to nothing about the strange experiments being conducted deep in the mosquito-clouded woods of northern Ontario, and when the ELA registered on its radar at all, it was as a baffling, wasteful nuisance. "They've been a terrible department to work for," David Schindler said of the DFO. "It's like they lie awake nights figuring out ways to destroy their employees' morale."

Schindler first ran afoul of his DFO minders in the midst of the ELA's crowning triumph—its establishment of the cause of acid rain. As DFO officials deliberated about how to respond to the findings in the mid-1980s, Schindler published a paper in *Science* setting a target of 10–15 kilograms of sulphate emissions per hectare of lake surface to avoid the calamity he had witnessed at the ELA's Lake 223. The Canadian government, however, had decided somewhat arbitrarily to set a target of 20 kilograms. DFO bureaucrats defended the position with zeal, some even accusing Schindler of wilfully undermining the official position (as opposed to merely gathering the data and reporting the scientific conclusions they established). Outspoken and irascible, Schindler was reprimanded several times throughout the 1980s for voicing inconvenient truths in public. In 1989 he resigned his position at the ELA after twenty-one years as director. "That stupidity is one of the reasons I left the project," he told me.

In the years to follow, the ELA faced the government's chopping block several times. Schindler: "Usually during each one of these cycles when ELA was on the block, a combination of editorials by science reporters who got it and the outcry from Environment Canada and good reviews by the Auditors General convinced the DFO bureaucrats to save the project. Outside of that, they had absolutely no understanding of what ELA did or why it was relevant."

The most serious threat came in 1996, as Jean Chrétien's Liberals struggled to pull the country out of the worst recession in decades. Not surprisingly, the ELA had few defenders within the DFO, but outside intervention—including what the *Winnipeg Free Press* described at the time as "a storm of protest in the Liberal caucus"—kept it from being eliminated entirely. Its staff reduced by two-thirds, the program soldiered on to meet the axe of Stephen Harper's Bill C-38.

The consolidation of science and politics under the DFO in 1979 was equally fateful to Canada's cod fishery. Two years earlier, Canada had extended its coastal jurisdiction from 3 to 200 miles. The extension came at the peak of the greatest boom in cod fishing in the history of the Grand Banks. More than a thousand industrial-scale trawlers, among them some of the largest fishing boats ever built, plied the waters of the North Atlantic in the boom years, hauling in more northern cod from 1960 to 1975 than had been harvested in the first 250 years of the fishery combined (1500–1750). Canadian government officials had once boasted that the cod fishery couldn't be depleted "unless nature reversed," but they hadn't anticipated the ecological force wielded by industrial "factory freezer" trawlers pulling massive dragnets along the sea floor.

The industry continued to thrive throughout the early 1980s, with annual cod catches reaching a peak of around a quarter-million tonnes at mid-decade. Fisheries ministers and bureaucrats alike continued to speak of abundant stocks and brighter days, but by the mid-1980s, a handful of DFO scientists began to warn of unsustainable harvests and dwindling populations. At a meeting of the Canadian Atlantic Fisheries Scientific Advisory Committee (CAFSAC) in November 1986, DFO scientist George Winters presented a paper containing particularly troubling evidence. Inshore catches were in steep decline, he attested, concluding that the DFO had been routinely overestimating cod stocks since 1977. CAFSAC—ostensibly a scientific body, but beholden to managers and a minister with higher priorities than science—acknowledged Winters's findings but dismissed them on what was later determined in a paper published in 1997 by the NRC's *Canadian Journal of Fisheries and Aquatic Sciences* to be "clearly unscientific" grounds. Quotas stayed high through the rest of the 1980s. In 1990, on the eve of the cod fishery's collapse and the economically crippling moratorium, CAFSAC presented three options for the 1991 quota: 100,000 tonnes (which DFO scientists had determined was the necessary target under a "conservation" scenario), 150,000 tonnes, and 170,000 tonnes. The committee dismissed the conservation scenario, and so the bureaucrats debated the merits of the 150,000- and 170,000-tonne targets against the "status quo" catch limit of 200,000 tonnes. The quota for 1991 was set at 190,000 tonnes. In 1992 the Canadian cod fishery, the richest aquatic bounty in human history, closed forever as a major commercial fishery. Stocks had collapsed, necessitating a moratorium so brutal and divisive for Newfoundland fishermen that the door had to be barred at the press conference at which the Fisheries minister, John Crosbie, announced it. The promise was that the commercial fishery would reopen when stocks rebounded in a couple of years. They never have.

The collapse of Canada's cod fishery inspired much study, recrimination and finger-pointing. In the first years of the moratorium, DFO bureaucrats, in what appeared to be an attempt to absolve themselves of responsibility, blamed abnormally cold waters and voracious seals for the vanished fish. A handful of the scientists in their employ argued otherwise. In a 1994 paper, DFO scientists Ransom Myers and Jeff Hutchings showed conclusively that "overexploitation precipitated the commercial extinction of northern cod." The official response was to dispute the findings and attack the scientists. A DFO deputy minister described the 1997 paper on the "unscientific" meddling in cod quotas, for example, as "tabloid journalism." Hutchings and Myers had both left the department for academia by then, and Myers testified to a parliamentary committee in 1997 that DFO bureaucrats engaged in "the intimidation and control of science," both by suppressing scientific evidence that conflicted with policy and by silencing the scientists who produced it.

Hutchings, meanwhile, co-authored the 1997 paper—an examination of the scientific lessons from the cod fishery's collapse and the department's failure to heed the warnings of its scientists. Entitled "Is scientific inquiry incompatible with government information control?" the paper notes that "nonscientific influences" greatly reduced the dissemination of critical information within the DFO and that "bureaucratic interference in government fisheries science compromises the DFO's efforts to sustain fish stocks." Which is to say, in general terms, that when science is beholden to the short-term exigencies of politics, science inevitably suffers. "The present framework for linking fisheries science with fisheries management," the paper concludes, "has permitted, intentionally or unintentionally, a suppression of scientific uncertainty and a failure to document comprehensively legitimate differences in scientific opinion. We suggest that the conservation of natural resources is not facilitated by science integrated within a political body."

"The collapse of the cod fishery," Hutchings told me, "is a very good example of what happens when you prioritize economic development at pretty much all costs and diminish the value of scientific advice."

THE OBVIOUS LESSON of the cod fishery's collapse, then, is that sustainable development of Canada's natural resources requires the old Borden model, with government scientists working at arm's length to establish the facts of the situation and bureaucrats and lawmakers writing policy based on their conclusions. In a sense, the current Conservative government has chosen to stand the lesson on its head. Instead of freeing scientists to establish the best methods to conserve natural resources, the government has crippled their ability to gather and disseminate the data demonstrating the consequences of policy decisions. If conservation of natural resources is not well served by government scientists integrated within a

government department, then why not simply abandon conservation as a goal? If government scientists routinely produce data at odds with policy goals, then muzzle government
science. Eliminate the programs and organizations that produce the data, and silence the remaining scientists when their
conclusions present challenges to those policy goals. The
problem of the cod fishery, from this point of view, wasn't the
collapse but the fact that government scientists figured out
why it happened. The problem wasn't that the DFO ignored
warnings; it was that the warnings were ever sounded in the
first place. Stephen Harper's Cabinet loves to talk about the
efficiencies of a streamlined policy, and what could be more
efficient than a government free of the troublesome friction
of data? Once again, the standard Harper agenda logic is at
work: if word of a smoking gun gets in the way of a lucrative
policy, especially in the resource extraction business, then
shoot the messenger.

This inversion did not happen all at once, of course. Canadians remain for the most part convinced of the value of
empirical data and the superiority of science-driven policy-
making. The shift to the post-Enlightenment darkness of evi-
dence-free policy has been slow and erratic, full of false starts
and oddly timed reversals and white noise. The Environment
Canada portfolio under seven years of Prime Minister Harper
provides perhaps the strongest case in point—an illustration
of the series of shrugs and sneers, afterthoughts and degra-
dations and wilful diminutions by which a once-vital Cabinet position was reduced to apologist-in-chief for unfettered
resource extraction.

When Harper's first minority government took power in
2006, the Environment portfolio was handed out almost as
an afterthought to Rona Ambrose, an MP with less than three

years' experience in Parliament and nothing in her past to suggest any knowledge of or interest in environmental issues (or science in general). Either Harper valued the department so little that he assumed the rest of the country and the world felt the same, or else he began his devaluation of the department almost too ambitiously, too pointedly, by handing it to one of the most junior members of his Cabinet.

Whether the motivation was malice or mere neglect (or both), Ambrose's appointment was an unmitigated disaster and an international embarrassment, neither of which a vulnerable minority government could afford. Handed a tangled mess of an environmental plan, Ambrose appeared unable to understand it or to sell it to the rest of the House or Canadians in general, at a time when concern about environmental issues—climate change, in particular—was at the top of the public priority list in opinion polls. Ambrose appeared ill-prepared at committee hearings—a post on the Green Party's website counted eight substantial factual errors in a single piece of testimony, and on one occasion an assistant deputy minister had to correct her in the middle of a presentation. And she was just as far out of her depth at international summits. At a November 2006 meeting of Environment ministers in Kenya, she engaged in a pointless war of words with France over its government's praise for Quebec's strong provincial record on environmental stewardship. Earlier in the year, Ambrose cancelled an interview with *National Post* columnist Don Martin at the last minute, after her office tried and failed to secure a promise that he would not ask about greenhouse gas emissions or the Kyoto treaty. "That's like promising the Health minister not to discuss the Canada Health Act or the Justice minister there'd be no talk of the Criminal Code," Martin wrote. "The big, if not only, question confronting Ambrose's portfolio in the months to come is simply Kyoto: Dead or alive?"

In early January 2007, after less than a year at the post, Ambrose was shuffled out of the job. In a *Globe and Mail* feature later that year, several members of Ambrose's inner circle of confidants told Jane Taber they blamed the PMO for Ambrose's flailing performance. "They accuse the office not only of failing to support her," Taber wrote, "but also of having no clue how to proceed with an environmental plan."

In the shuffle, Harper appointed John Baird—a combative politician and close ally—as minister of the Environment. "We recognize we need to do more on the environment," Harper told reporters. Baird brought bold statements and his brawling style to the job; he'd clearly been given the task of re-establishing the government's credibility on a key portfolio, as Harper manoeuvred into a more moderate position ahead of an election call that could come at any moment. At an international climate conference in Germany just three months into his tenure, Baird promised a greenhouse gas reduction strategy "among the toughest in the world," in avowed pursuit of Kyoto targets. He bashed the Liberal record relentlessly, taking full advantage of the substantial gap between the government's lofty talk and sparse action on climate change under Jean Chrétien. "We need to turn the corner," Baird said again and again. He also pledged to make Canada a world leader in "new green technologies." At a press conference one year after taking over the portfolio, Baird championed a new report from the National Roundtable on the Environment and the Economy (NRTEE) that urged the government to pursue an aggressive transition to a low-carbon economy. "Our government recognizes that climate change is one of the greatest challenges facing the world today," Baird announced, "and we have demonstrated leadership by taking real action to tackle this issue—many of which [*sic*] have been now recommended in the NRTEE's report. We agree that we must work in concert with the world, that policy certainty beyond the short-term is central, that technology deployment is imperative, and that an integrated approach to climate change and air pollution should be pursued."

Long gone were the uncertainties and equivocations of Ambrose's humiliating tenure. Baird was a crusading green champion, his ministry positioned as a much-needed corrective to a generation of Liberal foot-dragging. If this stance seems unrecognizable today, with the NRTEE a prominent casualty of Bill C-38 and John Baird about as likely to express concern for climate change as praise for an NDP policy initiative, it speaks to the degree to which Harper was caught flatfooted out of the gate on the Environment portfolio. In a minority government, the Conservatives desperately needed to convince Canadians that they shared their values and concerns, that they placed the same premium on environmental stewardship and held the same immutable belief in the primacy of science, especially on matters of ecological health, clean air and climate change.

The Conservatives won a second minority in October 2008, and they returned to Parliament with a new Environment minister who was in many respects an even stronger figure than Baird. Jim Prentice was a stalwart of the party's old Progressive Conservative faction and a vital ally who had helped broker the PC-Reform merger; he'd long been rumoured to be a top candidate for successor to Stephen Harper as party leader. Prentice was a loyalist but certainly not anyone's prop. He had policy ideas of his own and had no doubt about his own convictions. Prentice's appointment to the Environment portfolio suggests Harper knew he needed a strong voice in the position, someone who could continue to guard his minority government against accusations of neglect even as it negotiated a bold redirection of environmental policy that would pivot on the abandonment of its Kyoto commitments. Or perhaps he needed someone strong enough to sell that redirection but not strong enough to transcend it, someone who would win the battle but fall in the fight. Or perhaps it was both.

In any case, Jim Prentice was a forceful and polarizing figure in the role. The signature episode of his tenure as Environment minister was his attendance at COP15, the UN Climate Change Conference in Copenhagen in December 2009, at which the successor to the Kyoto Protocol was being negotiated under a global media spotlight. The Canadian government's delegation came to the conference with no intention of signing on to a new global treaty. Prentice's job was a brutal one—to hold the line on the government's rhetoric about strong action on climate change while essentially thumbing Canada's collective nose at the entire global environmental community, sneering at long-time allies behind a pleasant smile, pledging solidarity to the greater cause while blaming the United States and China for its own loss of resolve. And Prentice would have to accomplish this in openly hostile territory—the Canadian government was a pariah at COP15, condemned for derailing the negotiations and reviled for its fading commitment to the Kyoto treaty.

Even before he arrived, Prentice was the victim of a spoof that garnered international media attention. The notorious pranksters the Yes Men issued a false press release on Environment Canada letterhead pledging an ambitious increase in Canada's greenhouse gas reduction targets. The announcement was released through a fake Jim Prentice Twitter account and followed by a false retraction on a spoof Environment Canada website. The Climate Action Network also presented Prentice *in absentia* with a Fossil of the Day award for his pre-conference assertion that Canada was unwilling to be swayed from its position by "the Copenhagen hype." Before he'd even set foot in Denmark, Prentice was already a laughingstock in some circles.

There was little in Prentice's official statement at COP15 to drown out the laughter. The speech was barely three minutes long and contained little beyond toothless platitudes. The Canadian government was at COP15 "to secure a fair and effective and comprehensive climate change agreement." Canada endorsed "sustainable low-carbon economic growth." The "importance of our energy sector for meeting global demand," however, should not be overlooked. Prentice delivered the canned statement in the flat, for-the-record tones of a weather report. He often read directly from the paper on the podium in front of him, his eyes down nearly as much as up. His voice cracked when he hit the passage about the necessity of U.S. co-operation in any future climate deal. He looked and sounded like a headstrong student forced to read an unloved and meaningless passage against his will before an unforgiving audience.

A few months before Prentice's rote recitation of Harper loyalist cant at COP15, he had taken a much less malleable stance in meetings with Alberta premier Ed Stelmach and Environment minister Rob Renner. As revealed in briefing notes obtained by the Canadian Press long after the discussion, Prentice told his Alberta colleagues at the September 2009 meeting that "appropriate price signals" were essential to achieve the country's greenhouse gas emissions goals, advising them that the federal government intended to pursue "a carefully designed cap-and-trade system" to put a price on carbon. A few weeks later, on November 5, 2009, Prentice met with U.S. ambassador David Jacobson and told him that if Alberta and its oil industry did not adopt much tighter environmental regulations on their own, Prentice was prepared to "press federal environmental legislation" (this according to a cable sent by Jacobson to Washington recounting the meeting, which was later made public by WikiLeaks). Prentice also told Jacobson that on a recent trip to Norway he was amazed by the intensity of the vitriol toward Alberta's oil industry. The Canadian government, Prentice said, had been "too slow" to respond to the criticism and "failed to grasp the magnitude of the situation."

This was Jim Prentice's mood as he prepared for his appearance at COP15—troubled by Canada's growing reputation as an environmental laggard, reproachful of his home province and the government he represented for their intransigence on greenhouse gas emissions and stewardship, shocked at the toll it was all taking on Canada's once-sterling international reputation. Whatever his thoughts as he went through the motions at the conference hall in Copenhagen, he nevertheless played the part of the loyal soldier, revealing no misgivings, saying nothing about a price on carbon or the Canadian government's misapprehension of the scale and urgency of the climate crisis.

Once back in Canada, however, Prentice returned to combat mode. In front of a hometown business crowd in Calgary in early February 2010, Prentice voiced the same concerns about Canada's image that he had shared with the U.S. ambassador. "The development of the oil sands and the environmental footprint of these industrial activities have become an international issue, and as such, they now transcend the interests of any single corporation," Prentice said. "What is at issue on the international stage is our reputation as a country."

The crux of the speech is worth quoting at length; it is far and away the most openly critical statement any Cabinet minister under Stephen Harper has made about the business of resource extraction. Here's Prentice:

> It is no secret, and should be no surprise, that the general perception of the oil sands is profoundly negative. That is true both within Canada and internationally. We need to continue the positive work of industry, with investments in environmental technologies that will show the world how environmental responsibility and excellence can be taken to new levels. Absent this kind of Canadian leadership, we will be cast as a global poster child for environmentally unsound resource development. Canadians expect and deserve more than that. For those of you who doubt that the government of Canada lacks either the willingness or the authority to protect our national interests as a "clean energy superpower," think again. We do and we will... How we manage environmental issues post-Copenhagen will define Canada's future and our reputation on the international stage.

This was what Canada's Environment minister said as he stood before a podium in the elegant ballroom of the Palliser Hotel in downtown Calgary. He was just a few blocks from the southern border of his home riding, the tables before him filled with the city's business and political leaders, many of whom he had surely known all his professional life. It's hard to imagine a more familiar and friendly crowd for Jim Prentice outside his own living room.

Just six weeks earlier, Prentice had stood at another podium, in Copenhagen, and delivered his bland schoolboy statement on Canada's disregard for its Kyoto commitments, speaking on behalf of a government broadly understood in those quarters to be one of the most openly hostile in the world to the very idea of international action on climate change. And now, in this friendly room in Calgary, looking out over that sea of familiar faces, did Jim Prentice brag about the defiance he'd shown those oil-sands-hating environmentalists in Copenhagen? Did he crow about how his government was finally shaking off the shackles of Kyoto, a treaty so loathed in Alberta that it was listed alongside the hated gun registry and the Wheat Board on a popular bumper sticker as one of the three things that must be abolished to "Defend the West"? Did he play to the crowd, recite tried-and-true applause lines, bask in the friendly applause after the trials of COP15? No. Instead he argued that Canada's leadership on environmental issues was in danger of falling disastrously short, that his hometown's bread-and-butter industry was on the verge of turning the country into the world's paramount symbol of environmental destruction. He chastised, he challenged, he framed the discussion in the most dramatic terms possible. How the people in this lovely Palliser dining room chose to respond to the challenge of climate change—for what else could be meant by "post-Copenhagen" environmental issues?—would *define Canada's future.* In Calgary, with the memory of Copenhagen still large his mind, didn't Jim Prentice all but recant?

"It wasn't quite Daniel in the lions' den," *Maclean's* columnist Paul Wells wrote a few days later, "but it had a whiff of Nixon to China about it. Here was a senior Conservative cabinet minister putting the boots, at least rhetorically, to Alberta's oil sands." Wells noted that the tone shifted a couple of times throughout the speech, but in the end "Angry Jim" held sway in a speech remarkable for its pointed criticism. "It was unusually strong language coming from Prentice," Wells concluded. "Indeed it may even wind up meaning something significant."

A few months later, in June 2010, Jim Prentice travelled to Haida Gwaii, off the northwest coast of British Columbia, to dedicate a new National Marine Conservation Area. The new reserve was being hailed by the government and environmental groups alike as a landmark, because it protected both the waters that lapped the shores of Haida Gwaii and the land itself, recognizing the immutable interconnectedness of the whole ecosystem. Even more remarkable was Prentice's travelling companion—famed environmentalist and outspoken Harper critic David Suzuki. In a documentary that later aired on the CBC, Suzuki and Prentice explored the new marine park with Haida leader Guujaaw as their guide. As they sat on makeshift driftwood benches on the shores of Haida Gwaii, Suzuki chastised Prentice for the failures of COP15.

"We need a new approach," Prentice replied, toeing the government line. Suzuki conceded that environmentalists needed to learn to engage in less divisive dialogue. But corporate and political leaders, he argued, still lacked "an overriding sense of urgency." Prentice began to drift toward vague pronouncements about the passion of Canadians for "the outdoors."

Suzuki: "Please tell me that you understand that climate is an urgent issue. It can't be just . . ."

Prentice: "Oh, I fully appreciate it's an urgent issue."

Later in the summer, Prentice clowned for reporters in a braided toque at a photo op in the Arctic. "I've got the best job in the government," he said.

As the summer ended, Prentice reverted to stiff-lipped party-line statements as he prepared for the next round of UN climate talks in Cancún. In an interview with the Canadian Press in late September, he admitted that he was not optimistic that there would be a new international deal that year.

Just weeks later, on November 4, 2010, Jim Prentice abruptly resigned from Cabinet and from his seat in Parliament. Bruce Cheadle of the Canadian Press called the resignation an "astoundingly well-cloaked exit," noting that fellow Cabinet minister Tony Clement only learned about it on Twitter five minutes before it was announced. "Why Prentice is leaving now doesn't seem such a mystery," Cheadle wrote. His portfolio "is flat-lining."

"Stephen Harper has lost his best political fixer," was how columnist Don Martin summarized Prentice's departure. The job of Environment minister returned to the congenitally divisive John Baird, who barked defiantly about China's laggard progress on greenhouse gas emissions at December's Cancún climate talks while the world's environmentalists handed him five Fossil of the Day awards. The following month, inexperienced MP Peter Kent took over as Environment minister. And a few short months after that, in the spring of 2011, Stephen Harper finally won his majority and the Conservative government's war on science began in earnest.

By the time Bill C-38 passed a year later, the government's stance on science and environmental stewardship would be all but unrecognizable to Jim Prentice—let alone Robert Borden. As the leader of a minority government, Stephen Harper had dared only to chip away at the Enlightenment tradition of the arm's-length, evidence-based policy that had been Borden's great legacy. Once he had the "strong, stable majority" he'd sought for so long, however, Harper pounced on the opportunity to hack that legacy to shreds. The pieces worth keeping were to become the playthings of a centralized government in which all policy answered first to the immediate political goals of the government's front bench, and the bits that had proven especially difficult for the Harper minority would be tossed away forever. Good riddance to bad rubbish—and inconvenient science.

4

THE AGE OF WILFUL BLINDNESS

Science in the Harper Majority Years

MAY 2011–PRESENT

F OR MOST government departments—as for the country in general—the Harper agenda rolled out in the minority years much as it had for the Environment portfolio. It arrived in increments and lurches, bits and pieces. It took the form of cost-cutting and belt-tightening in turbulent economic times. It spoke of efficiencies and redundancies and smarter management, better business and common sense. Budget cuts, office closures and program overhauls were spaced out across the calendar, the changes all but patternless, minor waves in the greater heaving sea of government in the wake of the greatest economic crisis in nearly a century.

It would be an insult to the considerable political skill of Stephen Harper and his staff to assume this was anything other than intentional. Harper's inner circle surely understood they were breaking with traditions and toppling cherished institutions that had endured for generations. As Environment Canada staffers warned MP Michelle Rempel, parliamentary secretary to Environment minister Peter Kent, in briefing notes for a meeting with oil industry officials shortly before the 2012 omnibus budget bill was tabled, "The reforms, when introduced, may be very controversial." The agenda would succeed only if it remained diffuse, disjointed, largely invisible—if Canadians never caught a clear glimpse of the full panoramic scope of the shift, if they did not realize they were witnessing the disassembly of the very foundations of evidence-based policy-making. The agenda had to remain implicit and undefined.

Although program cuts and the muzzling of government scientists had been part of the Harper agenda since 2006, it was only after the majority was won in the spring of 2011 that the full thrust of its environmental policy became evident. After years of withering and whittling, after the embarrassment of Rona Ambrose and the firm backbone of Jim Prentice, the Conservative majority quickly erased all remaining doubt about the status of the Environment portfolio—as a barely necessary evil, a minor adjunct branch of Natural Resources Canada and a footnote to the serious business of industrial and economic policy. This shift in priorities was expressed most unequivocally in an open letter written by Natural Resources minister Joe Oliver and published in the *Globe and Mail* in early January 2012. Many mainstream environmental groups interpreted the letter as tantamount to a declaration of war.

The government's efforts to "diversify our markets" for Canada's natural resource exports, Oliver warned, were in danger of being blocked by "environmental and other radical groups." He accused this alleged cabal of being opposed to "any major project," from logging to hydroelectric dams, but the letter was widely interpreted as a warning shot ahead of the National Energy Board's Joint Review Panel hearings on the Northern Gateway pipeline. The new pipeline proposed to ship Alberta's bitumen across northern British Columbia to a new supertanker port at Kitimat—pretty much directly opposite the shores of Haida Gwaii, where Jim Prentice and David Suzuki had shared their thoughts on Canadian environmentalism so congenially less than two years earlier. "These groups threaten to hijack our regulatory system to achieve their radical ideological agenda," Oliver wrote. "They seek to exploit any loophole they can find, stacking public hearings with bodies to ensure that delays kill good projects. They use funding from foreign special interest groups to undermine Canada's national economic interest." The system of review for major resource projects, Oliver concluded, "is broken. It is time to take a look at it. It is an urgent matter of Canada's national interest."

Oliver's argument began from a verifiable fact: Canadian environmental and public interest groups had mounted campaigns urging citizens to sign up to testify before the Joint Review Panel on the Northern Gateway pipeline. A great many Canadians had answered the appeal. Although it's also true that some of the groups receive sizeable donations from American philanthropic organizations, these financial relationships were nothing new. And most important, they did not manufacture Canadian concern about the pipeline's environmental risks. Both industry and government had been arguing for months that the pipeline was a national infrastructure project of a scale equal to the St. Lawrence Seaway. The pipeline would cross 1,100 kilometres of wilderness and First Nations terrain, and it would create an oil-tanker shipping route that would introduce the spectre of an oil spill along a vast swath of British Columbia's coast. Did Oliver really find it surprising that Canadians were engaged in the process? It speaks volumes about his government's attitude toward the people it represents that the hundreds of Canadians who volunteered their time to participate in the process were apparently mere "bodies" to Oliver, faceless obstacles in the way of the incontrovertible Harper agenda.

Buried in the body of Oliver's histrionic letter, evidently without ironic intent, was a paean to cooler heads and rational decision-making. "Our regulatory system must be fair, independent, consider different viewpoints including those of Aboriginal communities, review the evidence dispassionately and then make an objective determination. It must be based on science and the facts." The timing of this appeal, as few outside the prime minister's inner circle knew at the time, only added to the irony. Oliver's letter went public just as the government's Group of Nine Cabinet ministers and other parliamentary insiders were concluding their review of government programs, which would trigger the omnibus budget bill's unprecedented assault on the whole Borden tradition of independent and objective policy development. The precise tools needed to review the environmental impact of the pipeline and insure its safety if and when it was built—environmental impact assessments, fish habitat experts, monitoring stations, and emergency response teams in northern B.C.—were destined for C-38 cuts. Oliver was grandstanding in defence of the very regulatory system his colleagues had just finished dismantling, asking for exactly the sort of objective review they had neglected to provide.

The Group of Nine's distinctive disposition toward expertise—scientific and otherwise—has become a signature of the

Harper agenda in general. Oliver appealed rhetorically to science and facts, dispassionate analysis and unconventional viewpoints, but the Harper agenda's 2012 cuts were formulated not by employing the rational calculations of the government's own experts or senior managers in each affected department or program but rather by consulting with the most self-interested and subjective actors in the whole process—executives from the industries whose regulation was under review. In a December 2011 joint letter, for example, four major oil-and-gas lobby groups wrote to Peter Kent and Joe Oliver recommending changes to six key pieces of legislation—the Fisheries Act, the National Energy Board Act, the Canadian Environmental Assessment Act, the Species at Risk Act, the Migratory Birds Convention Act, and the Navigable Waters Protection Act—to improve the regulatory process for their industry. And at a meeting with deputy International Trade minister Louis Lévesque in October, representatives from the Canadian Energy Pipeline Association also pushed for changes to the Navigable Waters Protection Act. In briefing notes for an Environment Canada official attending an oil-industry function in early 2012, government officials further acknowledged that the Canadian Association of Petroleum Producers (CAPP) had urged omnibus legislation—which would severely reduce the amount of time Canadians would be given to consider the changes and for Parliament to debate them—instead of piecemeal changes to the federal regulatory regime governing its industry.

When documents obtained by journalists under Access to Information laws brought these deliberations to light, government spokespeople acknowledged that officials at both Environment Canada and Natural Resources Canada had met with energy industry officials throughout the fall of 2011 but

insisted these were routine meetings. Joe Oliver's spokesperson also noted that the Natural Resources minister had met
with officials from both Greenpeace and the David Suzuki
Foundation. In any case, the regulatory apparatus governing
the energy industry was radically overhauled in 2012 through
a pair of omnibus bills that specifically addressed all six acts
that industry officials had identified as problems. The industry
got exactly what it wanted. Advocates for stronger environmental stewardship were outraged, seeing nothing in the budget bills to dissuade them from their interpretation of Oliver's
open letter. There would be no more walks on the beach with
David Suzuki. As a majority, the Conservative government
was flat-out hostile to environmental advocacy, contemptuous of environmental science, and indifferent to the advice of
experts past and present on the subject of stewardship.

"This is the most anti-environmental legislation we've seen
in decades," Rick Smith, executive director of the advocacy
group Environmental Defence, told the *Washington Post* after
the first omnibus bill passed. "Very clearly, a lot of these
changes are designed to expedite inappropriate pipeline pro-
posals. It's essentially a big gift to Big Oil."

For Smith, the omnibus budget bills represented a startling change in tone. His organization, one of the most moderate and influential in the Canadian environmental NGO
community, had worked amiably and effectively with the Conservatives as a minority government, helping to pass breakthrough legislation on environmental toxins—most notably,
a pre-emptive ban on the carcinogenic chemical Bisphenol A (BPA) in baby bottles even before Health Canada had
completed conclusive studies of its dangers to public health.
"We've done a lot of work with this government," Smith told
me in April 2012, a few weeks after Bill C-38 was first tabled.

"I find myself in the strange situation since January of having worked very closely with these guys and actually having achieved some significant progress on other files—namely, the regulation of toxic chemicals. And you know, since January, they have just been on a rampage with environmentalists, period. They just don't want to hear it when it comes to climate change science. In this latest budget, they've continued an undermining of the function of science in the federal government."

It's possible that Stephen Harper and his inner circle experienced some sort of radical shift in their thinking. There was, to be sure, a significant shift in tone and tactics. In the minority days, Jim Prentice had worked diligently with environmental groups to eliminate toxic chemicals. As a majority, however, the Conservative government treated its budget bill as a wish list for resource extraction industries and sent Joe Oliver to the pages of the *Globe and Mail* to declare war on foreign green radicals. So, yes, perhaps the PMO had changed direction for reasons that had nothing to do with retaining power as a vulnerable minority.

The more plausible explanation, however, is that the majority government's agenda had been the true Harper agenda all along. After all, the Conservatives' disposition toward environmental issues had verged on open contempt during Rona Ambrose's brief tenure on the portfolio, until they grasped the necessity of paying lip service to the public's green conscience. They'd even allowed a handful of substantial pieces of legislation to be passed, as long as they didn't have any direct impact on Canada's lucrative resource businesses; Bisphenol A is not a significant Canadian export commodity. But on core issues—Kyoto targets, most visibly, but even smaller resource extraction questions like the continuing export of carcinogenic chrysotile asbestos from Quebec—the Conservatives were often pointedly anti-environmental even as a minority.

When the full reins of government were finally theirs, they embarked on a wholesale evisceration of Canada's environmental stewardship regime. They rewrote legislation to suit oil lobbyists and pipeline developers. They ignored expert opinion both inside and outside their government when it argued against the omnibus bills, whether the dissenting opinion came from Greenpeace rabble-rousers or former PC Fisheries ministers. They demonized anyone who dared oppose them, even handing $8 million to Revenue Canada for the expressed purpose of auditing the organizations allegedly financed by those "foreign radicals" they insisted stood alone in staunch opposition to their agenda.

Even the warnings of long-time allies and the conservative movement's forefathers made no discernible impact on the Harper agenda once he had his vaunted strong, stable majority. The founder of the prime minister's party and the last Conservative to hold the job both urged a more moderate approach to stewardship. Harper ignored them both. "Early on in the prime minister's tenure, both Preston Manning and Brian Mulroney made the case that the greener path was also the safer path to implementing a resources-based economic agenda," Chantal Hébert wrote in the *Toronto Star*. "Manning has long argued that a Conservative movement determined to carve out a central place on the political landscape should take steps to own the environment issue. As for Mulroney, he had had an early taste of the future dynamics of the global energy/environment debate at the time of the controversy over Quebec's Great Whale hydro development." And though Hébert doesn't mention it, Mulroney, as previously discussed, had also presided over the most aggressive and effective environmental stewardship era in Canadian political history—one that welcomed opposition MPs at climate confabs in Geneva and employed the future leader of the Green Party in the senior ranks of its Environment ministry. But neither Manning's nor Mulroney's lessons were heeded, and the ominous prophecy Jim Prentice delivered to Calgary's business elite came to pass under the government he'd quit: *We will be cast as a global poster child for environmentally unsound resource development.*

EVERY ISSUE BECAME a battlefield in the majority years of the Harper agenda. Every expert was now suspect. Staffers in Joe Oliver's Natural Resources ministry warned him in briefing notes that the unimpeded growth of the oil sands and the toll it was taking on the boreal forest and the chemistry of the earth's atmosphere "has become a threat to Canada's international brand." Oliver never passed the warning on to Canadians. Environment Canada managers filled a thirty-three-page slideshow with "useful... concrete examples" of climate change's impact on Canada, none of which Peter Kent shared with the general public. The Food Expert Advisory Committee, convened by Health Canada, recommended that the government resume monitoring trans-fat levels in processed foods and consider regulations if the industry was unwilling to reduce those levels voluntarily. Health Minister Leona Aglukkaq ignored the recommendations. The scientists who conducted government-funded studies of ozone layer destruction and salmon depletion were forbidden to speak to the press or the public about what they had discovered. The NRTEE reported on the wisdom and efficiency of putting a price on carbon dioxide emissions (a recommendation also endorsed by 250 of the country's economics professors), so the government axed the NRTEE. During question period on May 15, 2012, Foreign Affairs minister John Baird said of the NRTEE, "Why should taxpayers have to pay for more than ten reports promoting a carbon tax, something that the people of Canada have repeatedly rejected?" Peter Kent explained that the NRTEE's analysis and expertise were "no longer required," that it was a relic from "before the internet, when there were few such sources of domestic, independent research and analysis on sustainable development." Shortly before the NRTEE finally disbanded, in early 2013, the government locked down its website and all its files, ensuring that its reports would not find too wide an audience. When Dalhousie's Jeff Hutchings, an international expert on fisheries, gave testimony before a parliamentary committee on the impact of climate change and aquaculture on biodiversity in Canada's coastal waters, Conservative MP Robert Sopuck told him, "In my view, scientists should stick to science."

After authorities foiled an alleged terrorist plot to derail a Via Rail train in April 2013, Stephen Harper told reporters he thought it would do no good for anyone to "commit sociology" on the accused conspirators. In response to the implied insult to the social sciences in general, University of Toronto philosopher Joseph Heath assembled a concise breakdown of the many counter-factual assumptions in the Conservative omnibus crime bill for the *Ottawa Citizen*. Heath noted that the Harper agenda's discounting of objective data and expert analysis served explicitly political goals. "As Harper's former chief of staff Ian Brodie has explained, part of their strategy in this area was specifically to antagonize criminologists and other intellectuals, so that the Conservatives could position themselves as defenders of common sense. 'Politically,' he said, 'it helped us tremendously to be attacked by this coalition, so we never really had to engage in the question of what the evidence actually shows about various approaches to crime.' Given this strategy, it's not entirely surprising that the government should regard anyone who wants to bring data to bear on the question of criminal justice policy as an enemy combatant."

Not content simply to dismiss or ignore unwelcome scientific evidence, Harper and his inner circle have come to treat those who gather such information as a suspect political class. Knowledgeable experts are just special interests getting in the way of the Harper agenda, and truth itself has been pitted against the Conservative government's policy goals. Science and research and peer-reviewed conclusions are treated as enemies of common sense. "Hostility to expertise in all of its forms is the closest thing that Canadian conservatives have to a unifying ideology," Heath concluded.

In the majority years, internal criticism has been outlawed and external dissent has become treason. Backbench Conservative MPs rise in the House of Commons to recite by rote the talking points prepared for them by the PMO, repeat stock phrases, hoot and howl and cheer and jeer on command, and treat the reduction of Parliament to an absurd slapstick sideshow as their only serious work. This is the triumph of a politics of pure posture, where taking a stance and ridiculing those who oppose it are more important than a proposed law's contents and implications, where message discipline is the sole means and rhetorical point-scoring the only worthwhile end, where loyalty is the only real value and the gathering and maintenance of power the highest purpose of government.

When a video surfaced online in which Conservative MP David Wilks was seen sharing his constituents' concerns about Bill C-38 and even confessing that he would consider voting against it if a dozen of his colleagues did the same, the corrective hand of party loyalty was swift and stern. Within hours, Wilks had disowned every independent opinion and gesture in the video. "I support this bill, and the jobs and growth measures that it will bring for Canadians in Kootenay-Columbia and right across the country," he declared in a hastily posted statement on his website. In the ensuing weeks and months, not another word was heard from Wilks on C-38 or any other matter of public policy.

The Harper majority's posture toward outside critics, meanwhile, was firmly established in Joe Oliver's Manichean open letter to critics of new energy and resource extraction projects. On one side was the government's point of view and on the other an opposing viewpoint so antithetical to the country's national interest that it must surely be foreign. Observers—both Canadian and foreign—could be excused for assuming that the government believed anyone who had any concerns about climate change, healthy ecosystems, or environmental stewardship in general to be an avowed enemy, because by the time of Oliver's letter the Conservatives had expressed utter contempt for all of the above and continued to do so with mounting zeal. Even back in 2007, when minority status obliged certain symbolic gestures toward the value of environmental policy, the entire Cabinet and all but two members of the Conservative caucus skipped out on a congratulatory reception for the Canadian members of the IPCC, which won the Nobel Peace Prize that year for their work tracking the global impact of climate change. The government refused to let Environment Canada scientists speak freely about sensitive ecological issues, slashed treasured environmental research and monitoring programs, and reneged on Canada's Kyoto commitment.

And then came Oliver's provocation to everyone and anyone who might question the wisdom of laying a pipeline across 1,100 kilometres of pristine wilderness, the sovereign territory of more than forty First Nations, the spawning grounds of countless millions of wild salmon, and the prime habitat of the rare Kermode bear, ending at a supertanker port opening on treacherous waters at the heart of the continent's largest remaining stand of intact temperate rainforest. The Northern Gateway pipeline was a study in extreme ecological risk along every metre of its proposed route. Along the way, it touched on nearly every environmental issue ever held dear by the Canadian public, from the health of freshwater ecosystems to the protection of singular natural wonders to the preservation of habitat to the perils of climate change. And yet to dissent from the government's position—to question it in any substantial way—was to side with foreign radicals, to conspire to commit treason. And finally, just as this debate was beginning to reach the general public, the government smashed the foundations of Canada's environmental regulatory apparatus in a pair of outsized budget bills and seized every opportunity to rail in Parliament against the Official Opposition's "job-killing carbon tax," as if to suggest that the very notion of serious action on climate change was contemptible. Again, what else were Canadians to conclude from this but that the Harper agenda saw no circumstance in which environmental concerns would override economic opportunities? Perhaps even that taking environmental concerns seriously was suspect in and of itself?

Although the Harper majority's mocking disdain for criticism and dissenting outside opinion is widely in evidence, there's been no spectacle as absurd as its post-omnibus talking point about job-killing carbon taxes. The fact that the NDP had proposed not a carbon tax but the very same cap-and-trade scheme for carbon pricing that the Conservatives had called for themselves as a minority government was among its lesser offences against logic. When the Canadian Press asked the government to confirm that the Conservatives had recommended the same policy, the PMO's communications director made no attempt to argue otherwise. "That's the past," he replied, as if talking about some odd nugget of Borden trivia and not the repeatedly stated public position of Environment minister Jim Prentice throughout 2009. (John Baird had also endorsed the policy in no fewer than eight letters to the editor published in major Canadian newspapers in 2008.) Never mind what was said long ago; listen only to what's said right now, over and over, with the precision and frequency of a fast-food chain's slogan. *Jobkillingcarbontax.*

Day after day in the House of Commons during the fall of 2012, one Conservative MP after another rose, usually in response to another topic entirely, to decry the NDP's nonexistent job-killing carbon tax proposal and its wholly unsubstantiated $20-billion-plus price tag. When Agriculture minister Gerry Ritz took his turn with the phrase, he stood with his script in hand, read the statement from it by rote, and then winked across the aisle. (Within weeks of his clowning mockery of parliamentary discourse, Ritz would be defending himself against accusations of negligence as XL Foods' meat-processing plant in Brooks, Alberta, reported an outbreak of *E. coli* and implemented a massive recall.) Aaron Wherry of *Maclean's* tallied up all the mentions of the carbon tax by Conservative MPS on September 17, 2012, and found at least twelve delivered in the House, to the press, or to the public on social media. "Maybe the Conservatives think you're stupid," Wherry speculated. "Maybe, more charitably, they just think they're smarter than you. Or maybe they assume that you're cynical enough—or enough of you are cynical—about this stuff that they can safely carry on like this. Or maybe they're terribly confused themselves. But sitting at home you probably shouldn't be joining in the laughter. Because ultimately the joke is on you."

This was the apotheosis of the Harper agenda's politics of posture, its clearest articulation. Facts were irrelevant. Irrefutable evidence that the Conservatives were demonizing a policy they themselves had championed was part of the joke. Parliament was a sideshow tent, and this was the government's best shtick. Climate change was the premise for a grand gag, and the policy option most often recommended by economists to address it was the punchline. The Conservatives repeated the joke over and over, under the chuckling assumption that their supporters would come to think of it as a riff on a truism—that the NDP did in fact want to implement a carbon tax, that it would inevitably kill jobs and cost billions, and that in any case their opponents, or really any chump silly enough to think a price on greenhouse gas emissions had merit, wouldn't be able to do much to slow their relentless pursuit of their own agenda.

There was, however, a comeback to the Conservatives' taunting gag, a self-inflicted wound hidden in the farce. In their relentless mockery, the Conservatives so radically polarized the discussion of Canada's energy future and its current environmental stewardship regime that they provided a powerful rallying point for climate change advocates, anti-oil sands activists, and a wide range of other opponents of the Conservative government and its anti-regulation, science-averse approach to resource development. Keystone XL, the "slam dunk" pipeline project required to continue to expand

Alberta's oil sands business at breakneck speed, became the very thing Jim Prentice had warned against: a global poster child for irresponsible resource development. The proposed pipeline became shorthand for the global fossil fuel industry's disregard for the health of the planet and the stability of its climate. By winking cynically at every Eurogreenie and foreign Hollywood enviro-radical and eco-socialist and egghead ecologist in its line of sight, Harper and his cronies made themselves global pariahs. And their brazen contempt for all things green reached its crescendo right at the moment the Conservatives needed to convince the U.S. government and the rest of the world that they could be taken at their word on the safety of Canada's pipelines and the stewardship of its oil projects. They ruined their credibility as sound environmental stewards in a gleeful chorus of taunts just when it was most crucial for their agenda.

"The time has come for America to move on climate, it appears," wrote Postmedia News columnist Michael Den Tandt in early 2013, "and Alberta's oil sands make an excellent, highly visible kicking dummy. As Martha Stewart was to excesses in the stock market, so the Keystone XL pipeline may become in the climate wars: an abject lesson. For this we have the master tacticians in the Harper PMO, and to a lesser extent NDP leader Tom Mulcair, to thank. Their polarized politics, and lack of foresight, are primarily responsible for placing Canada in this box."

THERE IS A COST, it turns out, to investing so heavily in the politics of pure posture. If you mock the House you preside over, you can become part of the joke yourself. If you discount and denigrate the work of the government you run and diminish its everyday operations, you may reduce your ability to pursue your own goals as well as the ones you've gleefully abandoned. If you turn politics into a preening carnival of competing circus acts, you might find yourself treated like a clown.

There is, as well, a literal cost to the politics of pure posture. Beyond the talking-point stunts in the House of Commons, the Conservative government has run up bills like no other in Canadian history on branding and sloganeering in the national media. Since 2009, the government has spent $113 million on its ubiquitous Economic Action Plan (EAP) ads and billboards alone. (Put another way, the government could have guaranteed funding for the Experimental Lakes Area for at least ten years if it had simply put the EAP branding exercise on hold for just one year.) The Action Plan grandstand began as a response to the global economic meltdown in 2008, as the government spent lavishly on public works and treated the most mundane highway upgrades and infrastructure investments as branding exercises under the EAP's triple-arrowed banner. The idea, evidently, was to give Canadians the illusion of furious activity everywhere, of a government racing to steer the economy away from the catastrophes seen in the nightly dispatches from foreclosed houses in American suburbs and smoke-choked streets in Greece. Canadians responded favourably at first, handing the Conservatives their 2011 majority in large measure because of the perception that they had skillfully piloted the economy through treacherous waters and brought it to a safe and relatively prosperous harbour without any permanent damage.

Even in 2013, though, the slogan was unchanged and the song remained the same. Animated EAP arrows continued to vector busily across Canadian TV screens during the NHL playoffs and the government continued to rail against the enemies of its strong, stable majority, but there was little else for a reasonable Canadian to attach the banner to. If the action plan was so frenetic and the vision so steadfast and bold, where were the monuments to its soaring productivity? Was there any substance behind the slogan? Did the arrows actually point to anything at all? And if they didn't, might Canadians start to wonder what exactly their money was being spent on?

The news throughout 2012 and early 2013 was not kind to the Conservatives' self-image as responsible fiscal managers and expert promoters of Canadian commerce around the world. From the procurement boondoggle on the $46-billion F-35 fighter jet program to an anti-terrorism strategy that couldn't account for $3.1 billion in spending, Harper's majority seemed to have delivered excess and incompetence instead of the long-promised strong, stable leadership. Even the government's own top priority—the expansion of Alberta's oil sands—was beset by criticism and controversy on all sides. Canadians had long prided themselves on their image as careful environmental stewards and global leaders on environmental legislation, and now they were in danger of being rejected by their largest trading partner because the Harper agenda had so fully ruined their reputation on such matters. In the heartland of America—the world's second-largest source of greenhouse gas emissions—Canadian oil (and Canadian policy) was deemed too dirty.

As the debate about Keystone XL reached a fever pitch—and Alberta's premier and federal Cabinet ministers alike shuttled to Washington, D.C., to make their case and try to re-establish their waning influence—a minor gesture in the politics of posture revealed its mounting cumulative costs. The United Nations Convention to Combat Desertification, to which the Canadian government had been a party since it was drafted in 1994, was exactly the sort of high-minded international environmental initiative Canada had once led. Renewing the funding for Canada's participation in 2013 would have cost $283,000 for the next three years—or less annually than the cost of a single Economic Action Plan ad aired during a *Hockey Night in Canada* playoff broadcast. Instead, Canada summarily abandoned its commitment to the convention, becoming the only UN member nation on earth not party to it.

The prime minister quoted obscure statistics about the small fraction of the program's funding that went to operations, and Foreign Affairs minister John Baird blustered about his disdain for "bureaucracy and talkfests"—standard Harper agenda poses, the sort of thing that resonated powerfully with many Canadian voters in the lead-up to the majority victory. In 2013, though, this abdication of international responsibility joined an extensive list of elisions and embarrassments on the environmental and diplomatic fronts. Instead of bold leadership, Canada now offered only petty partisanship. Canada was the country that backed out of Kyoto and sent minders to watch its scientists at international conferences. It was the home of dirty oil and the origin of leaky pipelines. (Just days after the government backed out of the desertification treaty, more than 10,000 barrels of Alberta's bitumen spilled down the streets of Mayflower, Arkansas, and into the international news.) Canada now stood entirely alone against international efforts to reduce the vulnerability of hundreds of millions of the world's poorest people to drought and climate change. This was not resolute leadership; this was alarming isolation. As former UN ambassador Robert Fowler told the CBC, it was "a departure from global citizenship." And this was, moreover, the intangible cost of posture politics—the stance had left Canada all alone in the world, railing against demons no one else could see.

THE BLOW TO Canada's international credibility is a disaster in itself, but even more significant is the damage it has done to the government's ability to protect the public interest and fulfill its own responsibilities. Even if a government chooses not to value stewardship and regulation very highly—even if it believes its efforts to diminish them are the best thing for the economy—Canadians still expect their government to fulfill these functions. If a company wants approval to build a pipeline, the government needs to be able to properly assess the risks. If there's an oil spill, the government is expected to be able to clean it up. But as a result of the Harper agenda's war on science and its relentless assault on the agencies, programs and departments responsible for environmental stewardship, the Canadian government can no longer accomplish some of its own most vital tasks.

The ironies of this diminished ability would be comical if their potential impacts weren't so tragic. In September 2012, for example, Environment minister Peter Kent used the twenty-fifth anniversary of the signing of the Montreal Protocol to issue a grandstanding statement about the government's track record as "a world leader" in atmospheric ozone science. He bragged about how Canadian government scientists in the years since the protocol was signed had invented the UV Index and created critical instruments for measuring atmospheric ozone. He neglected to mention that under his watch just a few months earlier, the government had shuttered PEARL, the only research facility on earth using those sorts of instruments to measure ozone levels in the High Arctic. His nostalgic hit list also skipped over the Conservative government's refusal to allow Environment Canada scientist David Tarasick to talk to the public about his groundbreaking research, which had detected a hole in the Arctic ozone layer twice the size of Ontario.

In a similar vein, Health Minister Leona Aglukkaq took the podium at the International Polar Year conference in Montreal in April 2012—at which event Environment Canada scientists were being tailed by media minders—to tout her government's many new research projects in the Arctic. "We need to know more about what's happening in these regions because changes there will affect every part of this planet," Aglukkaq said. Weeks earlier, she'd been a loyal member of the caucus tabling the budget that would shut down the only research station in Canada's High Arctic for several months and leave it scrambling for new funding. The same budget gave Revenue Canada $8 million to audit environmental groups believed to be backed by deep-pocketed foreign radicals—a sum that could have kept PEARL open for five years all by itself. As noted earlier, Revenue Canada used the lion's share of the money to scour the books of nearly nine hundred organizations in the first year after it was received and uncovered a single laggard—an anti-nuclear group run by doctors. Of the ten Canadian charities that receive the most foreign money, meanwhile, only one is in any way active on environmental issues: Ducks Unlimited, which received $33 million from outside Canada in 2010 and which was invited by Stephen Harper to participate as a charter member on the government's Hunting and Angling Advisory Panel in October 2012. Say this for the Harper agenda—it generates irony in bulk.

If only the damage were limited to the hypocritical pronouncements of preening Cabinet ministers. Celebrating past scientific research while diminishing the government's capacity to produce new findings is certainly troubling, but a far graver problem is the government's diminished capacity for environmental monitoring and emergency response. Environment Canada's Environmental Emergency Program, for example, lost sixty employees in the C-38 round of cuts and was forced to close all six of its regional offices. The program is the federal government's first response unit for oil spills, and its entire operational capability now resides in two offices—one in Montreal and another in suburban Ottawa. In early 2013 the federal government's environmental commissioner announced that the government no longer had sufficient resources to deal with a serious oil spill. The commissioner resigned a short time later. Environment Canada was also obliged to summarily abandon nearly five hundred environmental impact assessments, and the DFO has informed the Joint Review Panel assessing the Northern Gateway pipeline that it no longer has the capability to properly evaluate the project's environmental risks.

The DFO is one of the agencies hardest hit by the Harper agenda. The overhaul of the Fisheries Act and the substantial cuts to the DFO's operational capacity are among the most consequential—and controversial—changes to Canadian government under Stephen Harper. (This is ironic when you recall that one of the most devastating regulatory errors in Canadian history—the collapse of the Atlantic cod fishery due to overfishing—happened on the DFO's watch in significant measure because scientific data were manipulated for political ends and scientists were made to serve political masters.) In light of the revenues and livelihoods lost and communities devastated throughout Atlantic Canada, the DFO should by rights be one of the last government departments subjected to reckless, politically motivated cuts to its scientific work. Instead, it has been hacked back in size and scope. Its mandate has been shrunk and its operational capacity whittled away on nearly every front. The omnibus bill's cuts rendered more than a thousand DFO jobs redundant or obsolete, with about four hundred staff permanently eliminated and dozens of offices closed.

Agencies and offices charged with monitoring and responding to oil-related environmental disasters were particularly hard hit. The marine contaminant group on Vancouver Island, a team designed to play a critical emergency role in oil spills on the west coast, has been disbanded entirely. Numerous field offices across the country—including those in Prince George and Smithers, B.C., nearest to the proposed Northern Gateway pipeline's route—were closed forever. Seven of the DFO's eleven libraries were shut down. Jobs in habitat management have been especially vulnerable to the reduced budget's exigencies. And the Conservative government slashed another $100 million from the DFO's budget in 2013. The DFO's ability to gather data, monitor the health of aquatic ecosystems, and assess the impact of industrial activity on oceans and waterways has been greatly reduced. The department has become a shadow of its former self, another lowly adjunct to the serious business of resource exploitation.

The evisceration of the DFO has attracted ferocious criticism and protest. The concerted campaign against the closure of the Experimental Lakes Area and the public rebuke from several former Fisheries ministers as Bill C-38 was first being debated have been the most visible dissenting opinions, but there are precious few defenders of the shrunken DFO outside Stephen Harper's caucus. A Canadian Journalists for Free Expression report gave the department its lowest grade for encouraging press freedom (an F, compared to a C-minus for the entire government). Dalhousie University information management professor Peter Wells called the department's library closures "information destruction unworthy of a democracy." In response to Harper's assertion that science would determine the fate of the Northern Gateway pipeline, retired DFO staffer Otto Langer—the former habitat assessment chief for British Columbia—told the CBC, "[Harper] says the science will make the decision. Well, he's basically disembowelled the science. It's a cruel hoax that they're pulling over on the public."

In addition to budget cuts and office closures, the revisions to the Fisheries Act in Bill C-38 provide an even more enduring legacy. The changes were subtle in language but enormous in ramifications. Before its C-38 rewrite, the Fisheries Act explicitly applied to all "fish habitat." Wherever there were fish, the act's oversight and protections held sway. But the new version, as previously noted, covers only "fish of economic, cultural or ecological value." Habitat is a much broader category than any particular species of fish, and the "value" of any given species or population under any of the listed criteria is highly subjective. "Who will take on the role of God and decide which fish are of value?" wrote Alberta Fish and Game Association president Conrad Fennema in a letter to the prime minister and other government officials protesting the change. "We suggest that all fish species are of value in the big picture as every species plays a role in the survival of the next one up the chain." More than anything, the change in language has shifted the burden of proof from industry to the public sphere. Where industry once had to demonstrate that its proposed activities would not cause undue harm to vital fish habitat, now the government—or the public—will be required to prove the value of a given population of fish.

In a letter of protest written on behalf of the Canadian Society for Ecology and Evolution, Dalhousie University's Jeff Hutchings raised grave concerns about another cunning shift in the language in the new Fisheries Act. Whereas the previous act had legislated a ban on any "harmful" alteration or destruction of fish habitat—a robust legal tool—the new version prohibits only activities that do "serious harm" to fish populations that are part of commercial or recreational fisheries. "This revision will remove habitat protection for most of Canada's freshwater fish," Hutchings wrote. "The revision will also impair Canada's ability to fulfill its legislated obligations to prevent the extinction of aquatic species." Hutchings noted that more than half of Canada's freshwater fish and fully 80 percent of species at risk of extinction would no longer be protected under the new Fisheries Act.

The revised Fisheries Act is a categorical departure from some of Canada's deepest traditions: the recognition of the inherent and immutable value of nature; the vital role of government in representing and defending the public good against the excesses of industry; the idea that the public good itself possesses a value beyond the near-term profit of any given commercial enterprise; and the baseline assumption that economy is an instrument in the service of the public interest rather than a higher goal to be protected against the needless intrusions of the public at large. For more than a century, the Fisheries Act had stood in defence of these traditions, asserting that industry, whatever its merits and needs, could not act with impunity in the public sphere; that it must first demonstrate unequivocally that its activities would not take too high a toll on the commonweal. In a few tiny deletions and insertions buried deep within a budget bill of a scope without precedent in Canadian history, the Harper agenda has inverted the entire equation. The commonweal, whatever its merits, must now demonstrate how its claim on the preservation of fish and their habitat supersedes industry's right to profit from despoiling them. What's more, the same budget bill and several other acts of Parliament have shuttered or shrunk many of the agencies responsible for gathering the data necessary to establish the health of those ecosystems and their value to the public, and the government has forbidden its public servants from telling Canadians what they have learned about those topics.

This is an entirely new way of thinking about the role of government and the balance of public and private interests in Canadian life. This is a new political age—the age of wilful blindness, in which government's aim is to reduce its own ability to see the true cost of its policies and in which facts contrary to those policies are overlooked, eliminated, clamped down, ignored. This is a breach of fundamental democratic principles, an attempt to step beyond the boundaries of the Enlightenment tradition that established government as an instrument of the public interest and charged it with protecting that interest by gathering the best available information and using it to draft laws that best protect and extend and enhance it. Recall how Jeff Hutchings concluded his brief speech on Parliament Hill a few months after the Fisheries Act's revisions were unveiled: "An iron curtain is being drawn by government between science and society. Closed curtains, especially those made of iron, make for very dark rooms." And recall as well the words of his colleague Diane Orihel that same day: "We mourn the blindfold of ignorance imposed upon our once-great country."

As former Conservative strategist Allan Gregg argued in his "1984 in 2012" lecture, "Evidence, facts and reason form the *sine qua non* of not only good policy, but good government." A 2012 article in *Scientific American* on the alarming rise of "anti-science beliefs" in the corridors of U.S. government outlined the origins of this democratic tradition. The article's author, Shawn Lawrence Otto, recounted the thinking of Thomas Jefferson, inspired by John Locke's "Essay Concerning Human Understanding." Particularly critical was Locke's assertion that knowledge and its application in government are ultimately derived from "observations of the physical world." As Otto put it, "It was this idea—that the world is knowable and that objective, empirical knowledge is the most equitable basis for public policy—that stood as Jefferson's foundational argument for democracy."

To abandon this tradition, a government need not denounce Locke or Jefferson or even Allan Gregg. Nor does it need to declare independence from empirical knowledge *in toto* and outlaw its dissemination beyond its halls. To abandon this tradition, as the Harper agenda has demonstrated over and over again, a government must simply stop gathering certain facts on its own and stop employing what facts it has as the basis for its policies. In the age of wilful blindness, evidence and science haven't ceased to be; they have merely ceased to be seen.

In his seminal essay "What Is Enlightenment?" Immanuel Kant warned of the high cost to be paid by abandoning the tradition of evidence-based policy-making. "An epoch cannot conclude a pact that will commit succeeding ages, prevent them from increasing their significant insights, purging themselves of errors, and generally progressing in enlightenment.

That would be a crime against human nature whose proper destiny lies precisely in such progress. Therefore, succeeding ages are fully entitled to repudiate such decisions as unauthorized and outrageous."

This was self-evident to Kant in 1784, as it had been to Jefferson in 1776 and Locke in 1689. Governments might come and go, but the baseline Enlightenment commitment to human progress—to the gathering, interpretation, dissemination, and use of knowledge in the pursuit of more equitable government and a higher quality of life—has been considered sacrosanct for as long as democracy has existed. However powerfully driven by ideology, faith, or a relentless thirst for power, every democratic government must ultimately answer to the same objective truths. U.S. senator Daniel Patrick Moynihan's axiom that "everyone is entitled to their own opinions, but they are not entitled to their own facts" was widely cited after journalist Ron Suskind published a story in the *New York Times Magazine* in 2004 about the ideological excesses of the George W. Bush administration. An unnamed senior aide dismissed Suskind as a member of the "reality-based community," beholden to an outmoded worldview in which "solutions emerge from your judicious study of discernible reality." In the Bush administration, the aide explained, "we create our own reality." And one way to do so, in the age of wilful blindness, is simply to omit those parts of discernible reality most likely to contradict the invented reality you intend to create.

THE HARPER AGENDA'S shift to the post-reality community is not strictly a matter of ideology or political advantage. It involves both, but it is a deeper and a more diffuse transformation, a function of a political culture fundamentally different from those of the governments that have preceded it. Not just the goals of policy but the whole underlying foundation of government and the acceptable uses of the political sphere are now in flux in Canada. It's a shift perhaps most visible in those halls of government far from the contentious front lines of climate science and environmental stewardship but still subject to the new culture's warped take on the nature and purpose of science.

Let's consider, in particular, the fate of the National Research Council (NRC) under the Harper agenda. The substantial reordering that was to eventually trickle down to nearly every lab in Canada's most prestigious government research organization began with the appointment of Gary Goodyear as minister of state for Science and Technology in 2008. Although not directly in charge of the NRC, Goodyear has been the most visible and vocal spokesperson for the agency's transformation. As the press and critics have repeatedly noted, Goodyear has never worked as a scientist or engineer and does not even hold an undergraduate degree in a conventional scientific field; he was a chiropractor before running for office. He made headlines shortly after his appointment to Cabinet in 2008 for hesitant and confusing statements he made in response to a question about whether he believed in evolution. He argued that inquiries about his religious beliefs were irrelevant to his government work before asserting that "we are evolving every year, every decade." As evidence, he cited examples from his chiropractic work—wearing high heels, walking on cement—which are far outside the standard Biology 101 examples of evolution in action.

Goodyear's tenure as the ranking Cabinet minister on the Science and Technology brief saw the appointment of a new director at the NRC in 2010: John McDougall, a former petroleum engineer who had overseen the redirection of the Alberta Research Council into the Technology Futures subsection of the business-focused Alberta Innovates agency. McDougall soon made it clear he intended to bring substantial changes to the NRC as well. In an email to all NRC staff in March 2011, he outlined a new strategy for the organization. In the memo's preamble, he reminisced about his decision years earlier to transform his family's century-old business back in Edmonton. "I realized that history is an anchor that ties us to the past rather than a sail that catches the wind to power us forward."

The NRC's new strategy, however, remained an unfocused collection of management clichés. "We intend NRC," McDougall wrote, "to be a purposeful outcome based organization—the best RTO [research and technology organization] in the world contributing to Canada's future economic success. For us, outcomes mean that the work we do is successfully deployed and used to benefit our customers and partners in industry and government."

If the language was vague, the impact on the internal operations of the NRC was rapid and concrete. To become a sail, the anchor found itself radically reforged. The agency, once divided into semi-autonomous institutes, was reorganized around a handful of "theme areas" (McDougall's phrase) to be addressed through a variety of programs, some of which were to be designated as "flagship programs." Much of the executive power that had once resided at the level of the institutes moved upward into the new director's office, while the former managers of the various institutes scrambled to turn their research projects into programs aimed at theme areas and worthy of flagship status.

In the meantime, the NRC was engulfed in the larger chaos of the Harper agenda's war on science. Jobs were cut hither and yon, and some researchers found themselves banned from attending conferences in their fields, even if they had been invited to present papers. In one embarrassing incident, the staff of the Winnipeg-based Institute for Biodiagnostics received congratulatory Tim Hortons gift cards in the amount of three dollars each on the last day of work for a number of employees who'd recently been laid off. I spoke to a long-time NRC employee, who insisted on anonymity to protect his job, who described this transition phase as "complete chaos." Among other things, he noted that the NRC's accounting department had proven unable to keep up with billing clients for software licensed from the NRC during the protracted reorganization, leaving revenues uncollected for months at a time.

In March 2012, as McDougall carried on with the dramatic restructuring of the agency, Gary Goodyear stepped up to clarify the loftier goals being pursued. Goodyear said he envisioned the NRC as a "concierge" to Canadian business. "It will be hopefully a one-stop, 1-800 'I have a solution for your business problem,'" Goodyear told the CBC. "It will be the powerhouse that takes the ideas from wherever they come from ... and literally pushes those ideas into the marketplace through our business communities, as well as respond to the needs of the business community by providing, for example, research capacity and solutions." A year later, Goodyear reiterated the new mandate, declaring the NRC "open for business" and arguing that by doing too much to "create knowledge and push the frontiers of understanding," the agency had swayed too far from the needs of industry.

This is the essence of the new culture of government under Stephen Harper. The purpose of research—of science generally—is to create economic opportunities for industry, and the purpose of government is to assist in that process in whatever way it can. The business of Canada, to steal a phrase from Calvin Coolidge, is simply business, and technological innovation is its propulsion system. Government's job is to deliver innovations like theatre tickets to the front desk at a posh hotel. Innovation—a beloved buzzword of Harper-agenda apparatchiks like Goodyear and McDougall—is best understood in this context not as a transformative force of social progress but as a new feature on an appliance. This is a profound misunderstanding of not just what innovation is but how and where and why it happens. It is a substantial misrepresentation of what the NRC is and what it spends most of its time doing. And it demonstrates that the Harper agenda's many oversights might be the result not just of a mean-spirited ideological crusade but also—even primarily—of a myopic culture unable to comprehend the extent of its own ignorance. The wilfully blind, in other words, don't always intend to break everything they knock over; that's just what happens all too often when you're wilfully blind.

At the core of this blindness is the Harper agenda's abiding mistrust of expertise and its contempt for any kind of science not being applied directly to an economic activity of immediate benefit to Canadian industry and self-evident appeal to Conservative voters. Remember what David Schindler said about the C-38 cuts: "The kindest thing I can say is that these people don't know enough about science to know the value of what they are cutting."

When I spoke to Schindler, he drew a sharp distinction between his sometimes-fractious relationships with previous governments as an outspoken scientist and his more alarming interactions with the Harper-era Conservatives. "I went to testify in the hearings on C-38 and saw these guys sitting there looking like they were hypnotized, having the best scientists and senior officials in their own party—ministers of Fisheries in the past and things—tell them they were making a big mistake. They said, 'Oh no, you're wrong. We're not weakening anything.' It was just chilling to watch that."

Schindler rejected the idea that his past work at the ELA had made it a specific target of the cuts. His current work at the University of Alberta, studying the impact of oil sands extraction on freshwater ecosystems, might make his ELA legacy a ripe target for vindictive liquidation, but Schindler suggested instead that a broader cultural shift was at play. "They're cutting a lot of things that don't do anything connected to oil and gas. I think the larger picture is things that might come up with data that will slow industrial development more generally. You know, I think oil and gas is their primary target, but I think it's part of a much bigger picture than that."

Even when the Harper agenda is at its most targeted, then, it demonstrates little understanding of the nature of the work that government does day to day and how it might matter. My anonymous source at the NRC, whom I'll call Andrew, put it this way: "McDougall has this illusion that an expert is an expert. In a weird way, he kind of overestimates this. He thinks he can move us around from one field to another, and all we need is a couple months' retraining, because we're so smart we can do anything. So there's contempt for science, yes, but there's also a sort of weird overestimation of the capabilities of individual scientists."

Andrew pointed out that the very framing of the discussion around the NRC's reorganization betrayed a lack of knowledge about what the agency does. McDougall and Goodyear like to talk about shifting resources from "discovery-driven" science and "basic research" to technological innovation and applied research of more immediate use to Canadian industry. Pure science, Andrew argued, has been "a tiny part" of the NRC's work for decades. "The interesting thing is, the effect of moving the debate to those terms is that the general public has this vague idea that McDougall is cutting back on very abstract, airy-fairy, I don't know what—quantum theory or something like that. And then they ignore what's really happening, which is that he's doing terrible things to applied research that helps Canadian industry."

The explanation, again, is more cultural than ideological, more incidental than diabolical. "There are quite right-wing people in the U.S. who see the value of the kind of work we do," Andrew told me. "And the supposedly right-wing Conservative government doesn't. And it's because it's a different kind of right wing. What I'm seeing is more like a small-town hardware merchant, not willing to invest in anything if it's not going to pay off in the next six months. This guy [McDougall], what he would really like to be doing is just renting out the wind tunnel by the airport in Ottawa that the NRC owns, and getting cash on the barrelhead ... He's a hardware merchant. He wants you to pay a buck for a bag of nails."

To extend the analogy, a hardware merchant is generally not interested in putting his profits back into the kind of risky, long-term R&D that results in real innovation. If he makes a mint on a new high-efficiency lawnmower one summer, he's not likely to invest heavily in open-ended research on the science of grass growth or blade aerodynamics. He certainly won't take cash off his barrelhead and hand it to a bunch of eggheads playing around with gears and electric circuits that just might lead to an emissions-free engine or a self-propelled guidance system. He doesn't even necessarily want to know much about how the machines he's selling right now operate; he just wants to order more. What colours are selling this year? Is there a cheaper model that's a bit easier on fuel? Great, he'll take two dozen.

This point cannot be overstated: it is functionally impossible for a research organization like the NRC to properly identify, address and meet the R&D needs of industry according to the timeframe of a 1-800-hotline scenario or even inside a single business cycle or within a few years. There are any number of reasons why. Government research bodies aren't especially good at assessing the near-term technology needs of business, for one thing. Even private-sector venture capital firms, for whom this sort of next-generation R&D is their entire raison d'être, fail to read the marketplace's next twists and turns correctly more than nine times out of ten. And more than this—as the NRC's own track record attests—genuine innovation yielding major long-term economic benefits cannot be generated on demand. Perhaps no Harper agenda initiative so fully reveals the government's willingness to act in the total absence of evidence or best practices as the retooling of the NRC does.

By way of illustration, let's consider Canada's robust computer graphics industry, a posterchild for real innovation with a direct and substantial tie to the NRC in what many consider to be its "pure research" heyday. The industry's story begins in the early 1970s, when cash was plentiful, the timeframes languid, and the thinking often blue-sky. And it centres on a hippie motorcycle nut named Bill Buxton, who was studying electronic music at Queen's University in 1971. A professor tipped him off about some wild "digital music machine" up at the NRC lab in Ottawa—the keyboard of an organ hooked up to a state-of-the-art computer with 24K of memory that filled an entire room. So Buxton rode his bike up to Ottawa and started playing around with the thing. He never did use the traditional keyboard much, but he was fascinated by a pair of chunky devices outfitted with buttons and scroll wheels that could be used for "bimanual input." He could make music simply by whirling a couple of dials and pressing the buttons in sequence.

These exotic new tools fascinated Buxton and came to form the core of his research, which soon strayed far away from digital music production. One of the devices was what we now call a mouse. Buxton's free-ranging investigations, meanwhile, provided the foundation for his work as the innovative chief scientist at Silicon Graphics Inc. and Alias|Wavefront, a pair of companies that pioneered the commercial development of 3D computer graphics. Among other accolades, Alias|Wavefront won an Oscar in 2003 for its trailblazing work on computer-generated imagery for movies. The company, headquartered in Toronto, is exactly the kind of dynamic enterprise that MPs like to visit for photo ops to demonstrate all the amazing things that happen when governments invest in technological research. And it is a veritable certainty that no desperate manager would have suggested to an "outcome-based" director that the NRC's digital music machine was a prime candidate for a flagship program in the theme area of next-generation computer software. That's simply not how innovation works.

"Innovation is not about alchemy," Buxton wrote in *Business Week* in 2008. "In fact, innovation is not about invention. An idea may well start with an invention, but the bulk of the work and creativity is in that idea's augmentation and refinement ... The heart of the innovation process has to do with prospecting, mining, refining, and goldsmithing. Knowing how and where to look and recognizing gold when you find it is just the start. The path from staking a claim to piling up gold bars is a long and arduous one."

Buxton has argued elsewhere that Canada never would have found a place in CGI's vanguard if not for the opportunities he was given at the NRC in the early 1970s. I'd argue that everything Gary Goodyear and John McDougall have done to make the NRC a better concierge to industry has pretty much guaranteed that there will be no Buxton-like innovations at the NRC under their tenure. Hardware merchants, after all, aren't big on letting hippies play around with the merchandise in their stockrooms. It's bad for business— today's business, at least.

This is the nature of everyday science and technology in the age of wilful blindness. Not every catastrophe resulting from the Harper agenda bears the marks of an ideological crusade against environmental stewardship. Sometimes the culprit is a sort of self-inflicted clumsiness. The hardware merchants, proud of their ignorance and confident that only they can see in the dark, misconstrue the work of scientists as a matter of course. Guided by a common-sense logic unfettered by the constraints of data and evidence, they rebrand a research institute as a concierge desk, pile money on whatever entertainment options look good just then—the shiny and new, the wood-hewing and water-drawing—and then brag about how much they've spent on science this fiscal year. You could almost forgive them for knowing not what they do—if the toll it's all taking on the country were not so enormous.

5

LOST IN
THE DARK

The View from the Museum

SPRING 2013

T HERE'S A routine sort of chaos to Canadian government in the age of wilful blindness. It's endemic,
seemingly inevitable—there's just so much being
shifted around and knocked over with so little in the way of
foresight. As a random example, a fuel storage tank for a gen-
erator at Environment Canada's Canada Centre for Inland
Waters in Burlington, Ontario, started leaking diesel in
early 2011. As Mike De Souza of Postmedia News found out
through an Access to Information request, an Environment
Canada inspector reported the leak in March, but it wasn't
fixed until after a warning letter was sent the following April.
To be clear, an Environment Canada inspector reported a
dangerous fuel spill at one of Environment Canada's own
facilities, and nothing was done in response to the problem
for more than a year. Small wonder that a report presented to

Parliament by the federal environment commissioner, Scott Vaughan, argued that Environment Canada appeared to lack the wherewithal to respond adequately to infractions of its regulations across the board.

Another routine insult to scientific tradition: in early 2013, the DFO presented Andreas Muenchow, an American academic working alongside Canadian government scientists on an Arctic research project, with an agreement that would have required all his research to remain confidential unless he first received written consent from the government to make it public. Arguing that the new agreement was far more restrictive than a previous agreement he had made with the Canadian government in 2003, Muenchow refused to sign and posted the offending text on his blog. "The new draft language is excessively restrictive," he wrote, "and potentially projects Canadian government control onto me and those I work for and with."

The government's relentless posture politics creates its own strains of chaos and absurdity. In the fall of 2011, for example, Parks Canada decided to organize a press conference to announce the creation of a new national park on Sable Island, off the coast of Nova Scotia. Parks Canada had spent a year negotiating the new designation with the government of Nova Scotia and decided to time the announcement to coincide with the centennial of the agency's birth. When Parks Canada officials sent their agenda to the Privy Council Office (PCO) in Ottawa for approval, however, it was torn to pieces by allies of the prime minister. In the final days before the event, PCO officials summarily deleted the Parks Canada logo from a backgrounder document, demanded that the three Parks Canada officials who planned to join Environment minister Peter Kent and National Defence minister Peter MacKay on the dais be removed, and requested that an "ugly" Parks Canada centennial banner be taken down. The banner stayed, but the CEO of Parks Canada watched from the audience as MacKay informed the press that "50 years of conservation efforts culminate today with the Harper Government's signing of this agreement."

In political circles, this kind of shameless credit theft is sometimes called "bigfooting," and it offers a tidy little postcard image of the Harper agenda: a government hell-bent on diminishing the effectiveness of its environmental agencies, its loyalists shoving underlings out of the way so that Cabinet ministers can bask in the reflected glow of a tradition they had nothing to do with creating and intend to dismantle as soon as possible. (Nearby Kejimkujik National Park would have to curtail its winter services the following year for lack of funding.) The Harper agenda is a ribbon-cutting for a monument soon to be shut down, an anniversary celebration for an institution recently reduced to rubble. The whole Economic Action Plan campaign is in some sense a protracted attempt to paint a Conservative logo on the sound fiscal management and fastidious banking regulations of Liberal prime minister Paul Martin—which Stephen Harper's Conservatives opposed when they were all first enacted.

Even the odd piece of good news in the Canadian government's scientific circles has a certain randomness to it in the Harper majority years. In late April 2013, for example, scientists at the Experimental Lakes Area received word that their extraordinary research lab would likely survive. Ontario premier Kathleen Wynne announced that her government would provide some "operational support," and the Winnipeg-based International Institute for Sustainable Development (IISD) stepped up to agree to manage the ELA. The announcement of the IISD's commitment came from its new director, Scott Vaughan, the recently resigned federal environment commissioner. The new deal for the ELA arrived well beyond the eleventh hour—DFO officials had already begun dismantling offices and living facilities at the lakes—and there was no indication that the federal government had made any serious effort to secure a new home for the ELA in advance or even that it had done much other than agree to terms after the Ontario government and the IISD stepped forward. Still, Conservative MP Dean Del Mastro could soon be found bragging in his hometown paper about his government's hard work. The MyKawartha.com report included Del Mastro's startling claim that the government had been "developing remediation plans to ensure that the facilities are in suitable condition for a new operator." The sound management of the facilities—including the remediation of experiments after completion—was a role ably filled for forty-five years by ELA scientists, and in fact the DFO's primary activity at the lake sites that spring had been to start removing doors and windows. Nevertheless, the ELA, left for dead in the name of the Harper agenda, had somehow been spared from closing up shop for good. It passed for a feel-good story in Stephen Harper's Canada in 2013—it was the least bad thing the federal government had done to environmental science in recent memory—so why not take a bow?

So yes: the ELA was saved. The federal government would no longer have to answer directly to an agency of its own on matters of aquatic chemistry, and industry's toll on freshwater ecosystems, but data would at least still be gathered and the science would continue to be advanced. But what had once been an elite institution, Canada's answer to the Large Hadron Collider, in the words of University of Ottawa biologist

Jules Blais—was now a poor relation, gone begging to provincial governments and NGOs. Which points, finally, to the full cost of the Harper agenda and the grim faces of the casualties behind the line items cut and acronyms retired. A war on science, after all, is ultimately a war on *scientists*. And in an age where NRC researchers need to show immediate commercial applications and grant-seeking academics see the lion's share of federal backing go to technology instead of basic science, where world-class research labs are dismantled on a whim and scientists on the government payroll must seek the approval of media spin doctors before making even the most mundane comments on their work, Canada has become a place where the best and brightest scientists are less and less likely to feel welcome. What hotshot researchers or top-tier lab rats would choose to make their names and build their careers in Canada today? Who would want to work in an environment so anxious and chaotic, under an authority so arbitrary, for a nation now so contemptuous of certain kinds of science that it seems to have all but reneged on its commitment to the Enlightenment itself?

This scenario is far from hypothetical. One of the immediate responses to the defunding of PEARL, for example, was the departure of prominent atmospheric physicist Ted Shepherd, who had been using data from the research station in his work, from the University of Toronto to the University of Reading in the UK. The Harper agenda suddenly rediscovered the value of PEARL a year after cutting its funding, but it was too late to keep Shepherd in Canada or to fix the damage to the country's beleaguered reputation among climate scientists worldwide. When budgets are cut and offices closed, the losses are not only physical but also intellectual. The nation's current brain trust dissipates and diffuses. Senior scientists leave the government to retire or find safer posts in academia, and they are not replaced. Word spreads through international networks that Canada has turned on its scientists.

Even scientists working at Canadian universities have seen some of their most critical funding shift away from basic long-term research. NSERC, for example, has long been a critical funding pipeline for academic scientists. Its budget shrank by 5 percent in the 2012 budget, which included a moratorium on the Major Research Support program. As a result, the Bamfield Marine Sciences Centre, a 43-year-old research station on Vancouver Island, lost the funding that shared its critical data on ocean conditions with researchers around the world. The observation post survives, but its role in the larger scientific project of understanding the world's oceans has vanished at a time of climatic crisis. As NSERC funding priorities have shifted to business-oriented research, Bamfield's formerly stellar international reputation—and its ability to attract world-class scientists—has been radically diminished.

The Harper agenda's open hostility to certain kinds of research, in short, has not only shuttered many important facilities but also created a lost generation for Canadian science. "I think one thing that's going to happen that's unseen is a lot of the best scientists are just giving up," ELA founder David Schindler told me. "Some of them will put in their time, some will take early retirement. But they want out of there. There are a lot of really good people who are in their sixties that due to past cutbacks they have not replaced with the equivalent people. And meanwhile their salary scale can't even compete any longer with the worst universities. That's been a slow slide over twenty years as well. So there will be a huge decrease in the available talent within federal departments, at least in environmental sciences. And there are a number of people that are getting rather panicked about that. I've had discussions with some of the people who recognize the role that's needed in monitoring the oil sands, for example, and they just don't see where the personnel are going to come from with the current group."

THERE IS A wager implicit in the Harper agenda's war on science. The Conservative government is betting that meticulously message-disciplined rhetorical attacks, carefully stage-managed public pronouncements, and glossy Economic Action Plan ads on every *Hockey Night in Canada* broadcast will distract the Canadian public from noticing the full toll of the evisceration of government science and environmental stewardship until it is too late to be undone. Cheer along with the Economic Action Plan, celebrate the great victory of the War of 1812, work for something called "the Harper Government" instead of the "civil service," and pay no mind to the sounds of commotion, the howls of protest, and the wrecking ball's ungainly swing.

The Harper agenda's wager is predicated on the idea that a lie repeated often enough becomes its own kind of truth, that a talking point recited with enough frequency will eventually obscure every contradictory fact and reasoned counterargument. The clear signals of the nation's best evidence will be erased entirely by the white noise of spin. This is the blunt logic of message discipline as a political tool. There is no climate crisis, only job-killing carbon taxes. There is no gutted National Research Council in disarray, only a sleek concierge desk overflowing with new gadgets for the nation's industries. Regulations are shackles on economic growth and nothing more. Nature is without value until it is converted by industry into a natural resource, at which point it becomes a vital component of the national interest. Canada is and ever shall be the same vast, eternally bounteous treasure trove that François Gravé spied from his ship on the St. Lawrence in 1603. Ignore the pretences that have since attached themselves to Champlain's little nation-building project. Hew wood, draw water. Never mind the rest.

The Canada to be won in this bet is not the same country it was when Stephen Harper moved into 24 Sussex Drive in 2006. Stephen Harper's Canada is a country that has become a global symbol of callous, profit-driven hostility to sound environmental stewardship, a pariah state that lays ruin to international climate talks and abandons its sworn commitment to greenhouse gas reductions and opposes any measure that appears to take the planet's ecological health seriously. Stephen Harper's Canada is a country, alone among the democratic nations, that bars its scientists from discussing their work in public and sends spin doctors to ensure that message discipline trumps scientific fact even at academic conferences. It is a country where environmental advocacy is foreign and dissenting opinion treated as treasonous, a country where scientists march in the streets to reassert the primacy of the scientific method and doctors need to disrupt press conferences if expert analysis is to be heard in discussions of health-care policy.

Stephen Harper's Canada is a country where the government's elite research institution is criticized for failing to perform to the standards of a 1-800 customer service line. It is a country where chiropractors and petroleum engineers dismiss scientific inquiry as pointless and deem industrial R&D to be as paramount. It is a country that safeguards against oil spills by crippling its emergency response network and manages fisheries by withering away the government's ability to properly study fish. It is a country where Arctic sovereignty is predicated on extinguishing the government's only light in the Arctic night, a country where the world's foremost aquatic research lab is a costly nuisance, a drag on industry, a barrier to economic growth.

Stephen Harper's Canada is a country where the highest public interest is always in industrial expansion and economic growth.

Stephen Harper's Canada is a country where policy determines the facts and evidence is shaped to fit political goals, where everything from the casual commentary of scientists to the substance of the Fisheries Act has been massaged to forward a political agenda whose only real vision is the expansion of the government's own power and the exploitation of the country's natural wealth.

Stephen Harper's Canada is a country where a malicious and transparent agenda has been pursued to reduce the government's ability to gather data about the natural world, to eliminate the organizations charged with producing and interpreting that data wherever possible, to cripple the remaining agencies that respond to problems arising from that data, and to forbid government employees from talking about the implications of that data publicly—all of it to rush the country headlong into an age of wilful blindness.

This is what Canada gets if the Harper agenda wins its bet. This is the payoff: a diminished Canada; a narrow, mean place; a country afraid of open-ended questions and speculative science. A country out of step with all of its traditional allies, isolated and anxious and addicted to a shallow and brittle prosperity. A country standing firmly against progress on the twenty-first century's defining issue, the great existential challenge of our age—the question of how to reconcile our voracious appetite for energy with the unsustainable toll it is taking on the earth's climate and the natural world.

ON A FINE morning in early May—the same day Gary Goodyear and John McDougall held a press conference in Ottawa to announce that the National Research Council had switched its focus to "the identified research needs of Canadian businesses"—I paid a visit to the Royal Ontario Museum in Toronto. At the main entrance, visitors are greeted by the soaring two-storey skeleton of a *Futalognkosaurus*, a dinosaur discovered in 2007. A duck-billed Hadrosaur fossil, uncovered in the Red Deer Valley in Alberta in 1921, watches over the concourse beyond. They are bookends, of a sort, for a golden age in Canadian science.

It was a weekday morning, and the grand museum's four floors of galleries and exhibits were occupied mainly by dense little clutches of schoolchildren on field trips. As they streamed around me in excited bursts through the First Peoples hall, I gawked at Sitting Bull's war shirt and beaded Ojibwe vests, wondered at the meaning of the horns on a Blackfoot bonnet of weasel skin. There was an Iroquois horn donated by a country doctor in the late 1800s, Lakota moccasins dating to 1880, a metre-long ceremonial pipe, possibly Sioux, collected on a Red River expedition in 1858, and a grand old Abenaki birchbark canoe from the earliest days of European contact.

I knew some of what I was examining had been gathered by the Royal Canadian Institute over the first half-century of its existence. It had been analyzed and catalogued by some of Canada's first pure scientists, assembled into one of the young country's first great troves of knowledge, and then donated to the museum in the early 1900s, when it outgrew the institute's modest premises. The museum's placards and information kiosks don't indicate which items came from the institute's initial collection, perhaps because it isn't thought to matter. The value of knowledge is intrinsic and perpetual, sure to be safeguarded by any public institution charged with its care. Isn't this how we gauge a civilization's health—by how carefully it guards its treasured knowledge and hard-won wisdom?

One floor up, in the Schad Gallery of Biodiversity, the theme turns from the richness of human civilization to the abundant gifts of nature. Amid reproductions of polar bears and sharks and an imposing white rhino that lived at the Toronto Zoo until its death in 2008, the placards tell a story of grave crisis. "Due to global warming," one reads, "scientists predict the Arctic could be ice-free in summer by 2030." Another adds, "The loss of sea ice due to climate change threatens the Polar Bear's ability to hunt seals."

"Extracting natural resources is threatening the Boreal Forest ecosystem."

"Over 250 million people are directly affected by desertification."

These statements are presented, in the tradition of museums everywhere, as incontrovertible facts. The reason schoolkids crowd around the exhibits, after all, is because the Royal Ontario Museum is understood to be a repository of the foundational elements of knowledge, the building blocks of science, the basics. Whatever a Canadian citizen will one day become, she must first know this.

The Harper agenda knows better than to challenge the museum's assertions head-on. The PMO doesn't deny the problem of desertification—it merely abandons the international process by which the UN proposes to do something about it. The Natural Resources minister doesn't contradict the facts of

habitat loss and the poisoning of the waters in the boreal forest of northern Alberta—he just assists his caucus in dismantling the regulatory apparatus that might reduce the damage.
Even the prime minister himself feels obliged to pay lip service to the scale of the crisis of climate change—"perhaps the
biggest threat to confront the future of humanity today," he
once said, when his government was still a minority—but
nothing in the freighted language convinces him to place it
in the same universe of urgency as the rapid expansion of the
nation's industrial capacity to extract and burn fossil fuels.

Language can be a powerful tool. It can illuminate or it
can obfuscate with equal effectiveness. *Perhaps* implies a quali-
fication—*perhaps not*—in a way that's not as easily found in
phrases like *is threatening* and *due to.* There are few terms as
readily shaped and reshaped as *valuable,* and the difference
between banning activities that *are harmful to fish habitat* and
those that *do serious harm to fish populations* seems incidental to
the casual observer but catastrophic to someone well versed
in the legal precedents in environmental law. The Canadian
Museum of Civilization across the river from Parliament is to
become the Canadian Museum of History under the Harper
agenda, and how much will that really matter to the field-
tripping kids tromping through?

Maybe it's pedantic to point out that *civilization* is a broad,
outward-looking term, global in scope and universal in application, an embrace of all that has happened in the human
project for more than ten thousand years, whereas history can
be the narrative of a single place at a particular time and tell
only one particular version of the story. In any case, one of
the most celebrated discoveries in the final months before the
Museum of Civilization started its rebranding exercise was

Arctic archaeology curator Pat Sutherland's discovery of arti-
facts on Baffin Island indicating that it had once been home
to a Norse settlement that appears to have carried on trade
with the long-vanished Dorset peoples for centuries before
the arrival of Columbus. Sutherland's blockbuster discovery
was celebrated in the pages of *National Geographic* and on CBC
TV's *The Nature of Things* in the fall of 2012. It fundamentally
shifted our understanding of Canada's pre-Columbian past,
adding a new chapter half a millennium long to the story of
European exploration and exchange in the New World.

In April 2012—even as the *Nature of Things* episode about
her work was being filmed—Pat Sutherland was summarily
dismissed from her position at the Museum of Civilization.
Sutherland has been silent in the press, and so has everyone else who might know why she was fired from her position at the very moment she was bringing the museum global
acclaim for an unprecedented breakthrough. Andrew Gregg,
who wrote and directed the CBC program about Sutherland's
discovery, told the *Ottawa Citizen* he suspected her dismissal
was connected to the name change. "It's a complete shift in
ideology," he said. "The narrative that's coming out through
this government and our institutions has no room for a new
story about the Norse."

Perhaps Sutherland's dismissal has nothing to do with her
discovery that Canadian history is more complicated than it
had long been assumed to be. Perhaps. The Harper agenda,
in any case, includes $28 million in War of 1812 commemo-
rations and more than a quarter-million more to find John
Franklin's corpse, wagering that if it tells the simpler version
of the story often and loud enough, Canadians won't care to
know all the facts of their past. Or their present.

I'D LIKE TO bet against the Harper agenda's wager. My money is on another grand Canadian narrative, one that puts openness and uncertainty and constant analysis and revision at its very core. The hardware merchants can have their narrow, simple story. I'm betting that Canadians will want to continue to strive to be more than a nation of hardware merchants. I'm confident in my wager because it's the one I've known best my whole life—not because of any particular inclination or bias, but because it's the story Canadians were telling each other for a century until the Harper agenda interrupted us. And because it's the best story we have to tell. Because it's true.

I was born on a military base on the Canadian prairies in 1973. I grew up surrounded by military hardware and clearly delineated ranks, red maple leaves, and a quiet, resolute sort of patriotism. The first I heard of environmentalism was when Greenpeace blockaded the road from CFB Cold Lake to the town of Grand Centre in northern Alberta in the mid-1980s, in protest against tests of the U.S. military's cruise missiles; the high school students got the day off because they were bussed off-base to school. I attended military base schools or Catholic ones for almost all of my childhood. I completed a year of a business degree and hated it and switched to history, where one of my first lecture courses spoke of the importance of fur and railroads and wheat to Canada's past. All of which is to say that I was twenty years old at the earliest before I gave any serious thought to ecology or politics or protest. I grew up as fully immersed in Canada's official narrative as a child can, and even I knew, well before the age of twenty, that environmental stewardship and a deep embrace of the open spirit of scientific inquiry were central to my country's identity. I knew this because it was 1993, and it was simply who we were as Canadians.

Yes, Canada was a mercantile colony, a trading post, a source of raw materials, a nation dependent on one of three outsized trading partners for as long as it has been called Canada. But along the way, it has become something else. It is also a nation of curious explorers and adaptable voyageurs, a nation that is learning to atone for the ruin it has visited on its indigenous populations and learning moreover to treasure the profound knowledge of the land these First Nations have preserved despite it all. Canada is Sandford Fleming and Banting and Best, Norman Bethune and Lester B. Pearson and David Suzuki. The greatest Canadian hero of my childhood was a kid who lost his leg to cancer and then ran across the country to raise money to fund scientific research. Canada is the birthplace of Greenpeace and the host country for the signing of the Montreal Protocol. Canada is the nation that had once thought itself so big and empty and wild it couldn't possibly exhaust its bounty, and then the cod fishery and the outsized forests of the Pacific coast taught us otherwise, and now we were working through the messy business of finding a sustainable way of hewing wood and drawing water, and we intended to teach the world the lessons we'd learned in stewardship along the way. It was never a surprise to learn that the environmental movement had been born here, that global conservation treaties had been drafted and signed here, that Canadians chaired Rio's Earth Summit and led ipcc working groups and founded the Resilience Alliance. Striking delicate balances and brokering fair deals and advocating for greater protections was simply what we did. The flag stitched to the backpack of every international traveller in my youth was homage to this story.

And if Canada was never as simple as its symbols or as virtuous as it pretended to be, those symbols nevertheless represented—they still represent—the nation's greatest aspirations and its most noble ambitions. This is not a country that takes pride in disrupting international climate conferences, which I'd wager is why Jim Prentice couldn't bring himself to brag about it when he got back to Calgary. This is not a country that understands itself to be the worst environmental criminal on the planet, which is why even oil-and-gas executives turn confused and combative at the accusation. This is not a country that muzzles its scientists or dismantles its research labs or places its finest scientific institutions at the full mercy of business. Which is why, I'm wagering, Canadians are not going to put up with much more of the Harper agenda. See it clearly, and it reveals a picture that doesn't look like Canada at all.

SELECTED CHRONOLOGY OF CANADIAN SCIENCE

▬▬▬▬▬▬

1608: Samuel de Champlain, a navigator, cartographer and explorer, founds Quebec, the first permanent settlement in New France.

1635: The Collège des Jésuites, the first institution of higher learning in the New World, established in Quebec.

1747–49: Under the governorship of Roland-Michel Barrin de La Galissonière, a peer of the Académie royale des sciences in Paris, the colony of New France experiences a brief tenure as a centre of Enlightenment scholarship.

1820S ONWARD: Recurring waves of Scottish immigration bring some of Europe's leading scientific and engineering minds to Canada.

1842: Geological Survey of Canada founded by William Logan.

1849: Royal Canadian Institute (RCI) founded by Sir Sandford Fleming and Kivas Tully.

1853: The British government abandons its magnetic observatory in Toronto; the RCI takes over its operation.

1879: The RCI hosts Sir Sandford Fleming's first public lecture on his concept of standard time.

1882: Royal Society of Canada established; in 1884, it hosts the first ever meeting of the British Association for the Advancement of Science outside the UK, in Montreal.

1886: Central Experimental Farm established in Ottawa.

1899: St. Andrews Biological Station, Canada's first marine research station, begins operation on a floating scow at St. Andrews, NB.

1911: Sir Robert Borden elected prime minister, ushering in an era of sweeping progressive reforms of Canada's civil service.

1916: Borden's wartime government creates the Honorary Advisory Council for Scientific and Industrial Research, which later becomes the National Research Council.

1918: Borden's Conservatives pass the Civil Service Act, enshrining the government's commitment to evidence-based policy-making based on expert work conducted by arm's-length federal agencies.

1968: Experimental Lakes Area (ELA) established in northern Ontario, under the guidance of the Fisheries Research Board.

1973–74: ELA research led by co-founder David Schindler establishes a causal link between industrial and agricultural phosphate runoff and eutrophication producing catastrophic algal blooms on the Great Lakes.

1976: Schindler and his ELA colleagues begin research on the causes of acid rain.

1979: Scientific and bureaucratic arms of Canada's federal fisheries regime united in a single new office, the Department of Fisheries and Oceans (DFO).

1985: A landmark paper in the journal *Science* establishes a causal link between sulphur dioxide emissions from coal-fired power plants and other industrial processes and acid rain.

1987: Forty-six countries meet in Montreal to sign an agreement to restrict the use of the chlorofluorocarbons (CFCs) causing ozone layer depletion. The agreement, later ratified by 151 other nations, becomes known as the Montreal Protocol.

1987: The UN's World Commission on Environment and Development publishes its landmark Brundtland report, coining the term "sustainable development" and initiating the process leading to the 1992 Earth Summit in Rio and the Kyoto Protocol to combat climate change.

1988: The Canadian government under Brian Mulroney convenes "Our Changing Atmosphere: Implications for Global Security" conference in Toronto—the first major international meeting on climate change and other issues raised by the Brundtland report.

1990: At the World Climate Conference in Geneva, a delegation with representatives from all three federal political parties works collaboratively on Canadian climate policy.

1991: International Joint Commission passes Air Quality Agreement (AKA the "acid rain treaty") between Canada and the United States to reduce the cross-border air pollution causing acid rain.

1992: Canada's Atlantic cod fishery, once the most abundant on earth, is closed under an emergency moratorium after stocks collapse in the wake of a quota set at nearly double the "conservation" scenario target.

1992: Polar Environment Atmospheric Research Laboratory (PEARL) opens on Ellesmere Island, 1,100 kilometres south of the North Pole.

JANUARY 2006: Conservative Party captures 124 seats in federal election, forming its first minority government under Prime Minister Stephen Harper. Harper names MP Rona Ambrose as his first Environment minister.

JANUARY 2007: Ambrose replaced by John Baird as Environment minister.

NOVEMBER 2007: Environment Canada issues a new communications protocol, obliging all departmental staff—including researchers and scientists—to direct all media queries to media minders and "respond with approved lines" when necessary.

OCTOBER 2008: After being elected to its second minority government, the Conservatives return to power with Jim Prentice as Environment minister.

DECEMBER 2009: UN Climate Change Conference (COP15) in Copenhagen. Canada's opposition to a new agreement to succeed the Kyoto Protocol is widely derided.

FEBRUARY 2010: At a business luncheon in Calgary, Prentice warns that Canada is in danger of being "cast as a global poster child for environmentally unsound resource development" due to its lax regulations on new development in the oil sands.

JUNE 2010: The Conservative government abolishes the mandatory long-form census.

JULY 2010: Conservative government policy on HIV/AIDS introduced with minuscule funding for the treatment programs recommended by public health experts. The government also declines to sign the UN's Vienna Declaration, which advocates "evidence-based approaches" to drug policy as part of the global fight to eradicate HIV/AIDS.

NOVEMBER 4, 2010: In a surprise announcement, Jim Prentice resigns from Cabinet and from his seat in Parliament. John Baird replaces Prentice as Environment minister.

DECEMBER 2010: At the Cancún climate conference, Baird is presented with a record five Fossil of the Day awards for Canada's intransigence and its intention to renege on its Kyoto Protocol commitment.

MAY 2011: The Conservatives win their first majority government under Stephen Harper. In the new Cabinet, MP Peter Kent stays on in the Environment portfolio he was given in January.

SEPTEMBER 2011: Ignoring advice from legal experts and the opposition of a range of law enforcement groups including the Canadian Bar Association and the Canadian Civil Liberties Association, the Conservative government tables its omnibus crime bill, which includes mandatory minimum sentencing, the elimination of conditional sentencing, and a range of other reforms. It becomes law in March 2012.

FALL 2011: Kent and Natural Resources minister Joe Oliver and their staff meet with oil industry officials and fossil fuel lobbyists, who recommend reforming several key pieces of environmental legislation in a single "omnibus" bill.

OCTOBER–DECEMBER 2011: The "Group of Nine"—an informal group of senior Cabinet ministers and other Conservative loyalists—meets privately a couple of nights each week to decide on cuts to the following year's federal budget.

DECEMBER 2011: Environment minister Peter Kent announces that Canda is formally withdrawing from the Kyoto Protocol.

JANUARY 2012: In an open letter published in several prominent newspapers, Natural Resources minister Joe Oliver derides the "radical ideological agenda" of "environmental and other radical groups" that use funds from "foreign special interest groups" to oppose the government's approach to resource development.

FEBRUARY 2012: At a meeting of the American Association for the Advancement of Science in Vancouver, an open letter signed by numerous attending organizations—including Canadian Journalists for Free Expression (CJFE) and the Canadian Science Writers' Association—is sent to the Prime Minister's Office, calling for "unfettered access" to Canadian government scientists.

MARCH 2012: The Conservative government tables Bill C-38, its first "omnibus budget bill," which revises a half-dozen key pieces of environmental legislation, rewrites the Fisheries Act, closes several major environmental research facilities, and reduces the government's ability to monitor and respond to environmental problems across the board.

APRIL 2012: Environment Canada scientists attend the International Polar Year conference in Montreal, accompanied by media handlers who intervene in all their informal communications.

MAY 17, 2012: Staff members at Winnipeg's Freshwater Institute are informed that the Experimental Lakes Area will be shut down due to budget cuts in Bill C-38.

JUNE 2012: After a marathon vote on more than 300 opposition amendments in the House of Commons, Bill C-38 becomes law.

JULY 10, 2012: The Death of Evidence march from the Ottawa Convention Centre to Parliament Hill draws a crowd of more than two thousand—including hundreds of scientists in lab coats—to protest against the Conservative government's science policies.

SEPTEMBER 17, 2012: In the House of Commons and other public communications, Conservative MPs make twelve references to the NDP's alleged "job-killing carbon tax" in a single day.

WINTER 2012–13: As a result of budget cuts to Parks Canada, numerous national parks nationwide are obliged to close welcome centres, reduce winter access, and rely on volunteers for snow removal on winter recreation trails.

As a result of budget cuts to the Canadian Foundation for Climate and Atmospheric Science, the High Arctic research station PEARL shuts down for the winter, creating a gap in its data for the first time in eight years.

MARCH 2013: The Conservative government unilaterally withdraws support from the UN's Convention to Combat

Desertification, becoming the only UN member nation not party to the agreement.

APRIL 2013: The International Institute for Sustainable Development agrees to take over management of the ELA, saving the facility from permanent closure.

MAY 2013: The first results of the voluntary National Household Survey—the government's replacement for the long-form census—are released with the strong caveat that the results are less reliable than those of the cancelled mandatory census.

MAY 2013: Funding restored for PEARL.

More than $16 million in funding provided for Natural Resources Canada to run advertisements in Canada and the United States touting Canada's record for "responsible resource development."

SOURCE NOTES

_____

Every effort has been made to verify and document sources, and links were accurate at time of editing, but URLs change and page numbers get mis-transcribed. Should any detail or data point in this book need further clarification not found in this document, please contact Chris Turner directly at waronscience@gmail.com.

CHAPTER 1: MARCH OF THE LAB COATS
The description of the origins, planning, and execution of the Death of Evidence march in this chapter is based on extensive interviews with key organizers and participants, including Katie Gibbs (September 17, 2012), Diane Orihel (October 9, 2012), and Jeff Hutchings (April 5, 2013).
Physical descriptions of the march and its participants are drawn from several online sources documenting the march:
Meagan Fitzpatrick (CBC). "Death of scientific evidence mourned on Parliament Hill," July 10, 2012. www.cbc.ca/news/politics/ story/2012/07/10/pol-death-evidence-protest-parliament-hill.html. This report includes CBC reporter Kady O'Malley's detailed live-blog of the march.
David Ljunggren (Reuters). "Canadian scientists protest against spending cuts," July 10, 2013. www.reuters.com/article/2012/ 07/10/canada-politics-science-idUSL2E8IA5CP20120710

John Hansen (YouTube video). www.youtube.com/watch?v=
    1aT5JZ-ppME
Full transcripts of all the march's Parliament Hill speeches have been
    archived online at The Tyee: thetyee.ca/Opinion/2012/07/16/
    Death-of-Evidence/
The description of the staff meeting at Winnipeg's Freshwater Institute
    at which the closure of the ELA was announced is drawn from the
    author's interview with Diane Orihel, October 9, 2012.
The Andrew Coyne quote on page 8 is from "Bill C-38 shows
    us how far Parliament has fallen," *National Post*, April 30,
    2012. http://fullcomment.nationalpost.com/2012/04/30/
    andrew-coyne-bill-c-38-shows-us-how-far-parliament-has-fallen/
The Cynthia Bragg quote on page 8 is from "Beware of what's beyond
    the rhetoric in the federal budget," *Guelph Mercury*, April 8, 2012.
    www.guelphmercury.com/opinion-story/2781438-beware-of-what-s-
    beyond-the-rhetoric-in-the-federal-budget/
A collection of international media stories about the Death of Evidence
    march can be found at www.deathofevidence.ca/media
The *Guardian* coverage cited is Alice Bell, "Why Canadian scientists
    need our support," *The Guardian*, July 11, 2012. www.guardian.co.uk/
    commentisfree/2012/jul/11/canada-scientists-strike-protests
The Christopher Hume quote on page 12 is from "Stephen Harper is
    blind to science," *Toronto Star*, July 13, 2012. www.thestar.com/news/
    gta/2012/07/13/christopher_hume_stephen_harper_is_blind_to_
    science.html
The Stephen Harper quote on page 13 was widely reported. See, for
    example, "Harper defends independence of pipeline approval
    process," CBC News, August 7, 2012. www.cbc.ca/news/business/
    story/2012/08/07/pol-gateway-tuesday-harper-bc.html

CHAPTER 2: LANDSCAPE AT TWILIGHT
The George Monbiot quote on page 18 is from "2012: the year we
    did our best to abandon the natural world," *The Guardian*, Decem-
    ber 31, 2012. www.guardian.co.uk/commentisfree/2012/dec/31/
    year-abandon-natural-world
Information on the state of the global climate was drawn from a range of
    sources. For temperature conditions in the United States and around
    the world, see:
Neela Banerjee (*Los Angeles Times*). "2012 was among the 10 hottest
    years on record globally," January 15, 2013. www.latimes.com/
    news/science/sciencenow/la-sci-sn-higher-global-tempera-
    tures-nasa-noaa-20120115,0,7943007.story

Justin Gillis (*New York Times*). "Not Even Close: 2012 Was Hottest Ever in U.S.," January 8, 2013. www.nytimes.com/2013/01/09/science/earth/2012-was-hottest-year-ever-in-us.html

For extreme Canadian climate and weather details, see:

CBC News/Canadian Press. "'Big heat' tops Canadian weather stories of 2012," December 20, 2012. www.cbc.ca/news/canada/story/2012/12/20/canada-top-weather-stories-2012.html

Niamh Scallan (*Toronto Star*). "Fruit industry in Ontario devastated by extreme weather," May 9, 2012. www.thestar.com/business/2012/05/09/fruit_industry_in_ontario_devastated_by_extreme_weather.html

Maclean's/Canadian Press. "Pine beetles so widespread they're contributing to climate change: study," November 25, 2012. www2.macleans.ca/2012/11/25/pine-beetles-so-widespread-theyre-contributing-to-climate-change-study/

Aaron Hinks (*Grande Prairie Herald Tribune*). "Pine beetles continue to devastate boreal forest," September 9, 2012. www.dailyherald tribune.com/2012/09/09/pine-beetles-continue-to-devastate-boreal-forest

Details on the 2012 Arctic sea ice melt (including the comparative size of the melted area) are from World Meteorological Organization (press release), "2012: Record Arctic Sea Ice Melt, Multiple Extremes and High Temperatures," November 28, 2012. www.wmo.int/pages/mediacentre/press_releases/pr_966_en.html

For Canadian stewardship rankings, see:

CBC News. "Canada last among G8 on climate change action: report," July 1, 2009. www.cbc.ca/news/technology/story/2009/07/01/tech-climate-scorecard-wwf.html

The Huffington Post Canada. "Canada's Environmental Health Lags Developed World: Conference Board Report," January 17, 2013. www.huffingtonpost.ca/2013/01/17/canada-environmental-health-ranking_n_2497459.html

For details on the XL Foods *E. coli* outbreak, see *Toronto Star*/Canadian Press, "XL Foods: Independent review blames lax attitudes for beef recall," June 5, 2013. www.thestar.com/news/canada/2013/06/05/xl_foods_independent_review_blames_lax_attitudes_for_beef_recall.html

For water conditions in First Nations communities, see âpihtawikosisân (blog post), "Dirty Water, Dirty Secret," November 8, 2012. http://apihtawikosisan.com/2012/11/08/dirty-water-dirty-secret-full-article/

Munir Sheikh's quote is from his essay "Good Data and Intelligent Government," *New Directions for Intelligent Government in Canada* (published by the Centre for the Study of Living Standards, available online at www.csls.ca/festschrift/Sheikh.pdf, p. 333).

Datalibre.ca's "CensusWatch" list is online at datalibre.ca/census-watch/

For details on the National Household Survey and Statistics Canada's data reliability warnings, see Rennie Steve's article (*Toronto Star*/ Canadian Press) "National Household Survey: Statistics Canada disclaimer warns of 'non-response error,'" May 8, 2013. www.thestar. com/news/canada/2013/05/08/national_household_survey_statistics_canada_disclaimer_warns_of_nonresponse_error.html

The quote from the Environment Canada's 2008 "media protocol" is from Pallab Ghosh's article (BBC News) "Canadian government is 'muzzling its scientists,'" February 17, 2012. www.bbc.co.uk/news/science-environment-16861468.

For details of Kristina Miller's silencing at the hands of the Privy Council Office, see Margaret Munro's article (Postmedia News) "Ottawa silences scientist over West Coast salmon study," *Vancouver Sun*, July 27, 2011. www.vancouversun.com/technology/Ottawa+silences+scientist+over+West+Coast+salmon+study/5162745/story.html

Mike De Souza's experiences with Environment Canada and David Tarasick were recounted by Kai Benson in "Silence of the labs," *Ryerson Review of Journalism*, January 2, 2013. http://www.rrj.ca/m25739/

Mike De Souza (Postmedia News) also wrote, "Scientist speaks out after finding 'record' ozone hole over Canadian Arctic," *National Post*, October 21, 2011. http://news.nationalpost.com/2011/10/21/scientist-speaks-out-after-finding-record-ozone-hole-over-canadian-arctic/

Environment Canada's media control tactics at the International Polar Year conference were reported by Margaret Munro (Postmedia News) in "Critics pan instructions to Environment Canada scientists at Montreal conference," April 23, 2012. www.canada.com/technology/national/6500175/story.html

The *Nature* quote is taken from the editorial "Frozen out," March 1, 2013. www.nature.com/nature/journal/v483/n7387/full/483006a.html

The quote from the AAAS meeting's open letter and the details of its drafting were taken from Petti Fong's article "Federal scientists say they're being muzzled," *Toronto Star*, February 17, 2012. www.thestar. com/news/canada/2012/02/17/federal_scientists_say_theyre_being_muzzled.html

The Stephen Strauss quotes are from the author's interview with Strauss, April 18, 2012.

Details of the federal government's AIDS policy and Julio Montaner's criticism of it (including the quoted phrase) can be found in

"Ottawa's HIV/AIDS funding disappoints some," CBC News, July 20, 2010. www.cbc.ca/news/health/story/2010/07/20/hiv-funding-vienna-declaration.html

There were numerous reports, responses and debunkings of the omnibus crime bill in the media in 2010. For a representative example, see Charles Pascal's editorial "Harper tough on crime but soft on facts," *Toronto Star*, November 17, 2010 (www.thestar.com/opinion/editorialopinion/2010/11/17/harper_tough_on_crime_but_soft_on_facts.html), which summarizes Paula Mallea's report "The Fear Factor: Stephen Harper's Tough on Crime Agenda" for the Canadian Centre for Policy Alternatives (full report available online at www.policyalternatives.ca/publications/reports/fear-factor).

For a recount and analysis of Stockwell Day's "unreported crime" press conference, see Kady O'Malley's report "Stockwell Day and the Mystery of the Unreported Crime Surveys," CBC News, August 3, 2010. www.cbc.ca/news/politics/inside-politics-blog/2010/08/stockwell-day-and-the-mystery-of-the-unreported-crime-surveys.html

For an overview of Bill C-38 with emphasis on its cuts to environmental science, see "Green gets mean with Ottawa," *Vancouver Sun*, June 8, 2012. www.canada.com/vancouversun/news/westcoastnews/story.html?id=1354a576-d17c-49eb-a88a-d14801cb5c19

See also:

Postmedia News. "Tories cutting vital climate science, critics say," *National Post,* September 14, 2011. http://news.nationalpost.com/2011/09/14/tories-cutting-vital-climate-science-critics-say/

Suzanne Goldenberg. "Canada's PM Harper faces revolt by scientists," *The Guardian*, July 10, 2012. www.guardian.co.uk/environment/2012/jul/09/canada-stephen-harper-revolt-scientists

Doug Cuthand. "Ottawa socking it to First Nations institutions," *Saskatoon StarPhoenix*, April 27, 2012. www2.canada.com/saskatoonstarphoenix/news/forum/story.html?id=9de5c65e-1f17-4fd4-91c5-f19d2102fa73

Allan Woods (Ottawa Bureau). "Conservative government shutting down world-class freshwater research facility in northern Ontario," *Toronto Star*, May 17, 2012. www.thestar.com/news/canada/2012/05/17/conservative_government_shutting_down_worldclass_freshwater_research_facility_in_northern_ontario.html

The Scott Vaughan quote on page 26 is from "Stephen Harper's environment watchdog to investigate 'risks' of federal budget bill" by Mike De Souza, Postmedia News, September 7, 2012. http://o.canada.com/2012/09/07/stephen-harpers-environment-watchdog-to-investigate-risks-of-federal-budget-bill/

Details on the restructuring of the National Research Council (including Gary Goodyear's "concierge" quote) can be found in "National
Research Council to 'refocus' to serve business," CBC News, March 6,
2012 (www.cbc.ca/news/technology/story/2012/03/06/technology-
goodyear-national-research-council.html). For further references and
sources, see Source Notes for Chapter 4.

For details of C-38's impact on environmental assessments, see Larry
Pynn, "Feds walk away from environmental assessments on almost
500 projects in B.C.," *Vancouver Sun*, August 22, 2012. www.vancou-
versun.com/technology/Federals+dump+environmental+assessments
+almost+projects/7125419/story.html

For details on the impact of Revenue Canada audits of environmental
groups, see Kate Webb, "One year and $5 million later, Harper's
charity crackdown nets just one bad egg," *Metro*, March 30, 2013.
http://metronews.ca/news/vancouver/613999/one-year-and-5-million-
later-harpers-charity-crackdown-nets-just-one-bad-egg/

For details on opposition to changes to the Fisheries Act, see "Fisheries changes attacked in prestigious Science journal," CBC News/
Canadian Press, June 22, 2012. www.cbc.ca/news/canada/
story/2012/06/22/pol-cp-fisheries-scientists-budget-bill-concerns.
html?cmp=rss

The David Schindler quote on page 29 is from "Canada stops funding famed experimental lakes science program" by Margaret
Munro, Postmedia News, May 17, 2012. www.vancouverdesi.com/
news/canada-stops-funding-famed-experimental-lakes-science-pro-
gram/140148/

Details on the "Group of Nine" and their work on the cuts contained in
Bill C-38 are drawn from "He's got the future of the PS in his hands"
by Jason Fekete, *Ottawa Citizen*, March 17, 2012 (www2.canada.com/
ottawacitizen/news/observer/story.html?id=fd79af66-4188-4ad2-
be76-d738451b3af9). The quote on page 29 from Tony Clement's
speech at the Manning Centre is drawn from the same *Citizen* story.

The quote from Christopher Plunkett on page 30 is drawn from "Canadian government overhauling environmental rules to aid oil extraction" by Juliet Eilperin, *Washington Post*, June 3, 2012. http://articles.
washingtonpost.com/2012-06-03/national/35460476_1_tar-sands-
canadian-government-macdonald-laurier-institute

The quotes on pages 32–33 from Allan Gregg's Carleton University
lecture "1984 in 2012 — The Assault on Reason" are taken from
the version of the text posted online at Gregg's blog (allangregg.
com/?p=80). The Gregg quote on page 33 is from his article "In

defence of reason," *Toronto Star*, October 8, 2012. www.thestar.com/
opinion/editorialopinion/2012/10/08/in_defence_of_reason.html
The Daniel Patrick Moynihan quote on page 34 has appeared in a number of different versions, sometimes attributed to one of several other
people. The version in this text is drawn from "'You're not entitled
to your own facts' vs. That's your opinion. Kiss my ad," a post at Jay
Rosen's PressThink, August 24, 2012. http://pressthink.org/2012/08/
youre-not-entitled-to-your-own-facts-vs-thats-your-opinion-kiss-my-ad/
Details on the closure of PEARL and quotes on page 36 are drawn from
the author's interview with James Drummond, PEARL's chief investigator, April 18, 2012. See also "High Arctic research station forced
to close," CBC News, February 28, 2012 (www.cbc.ca/news/politics/
story/2012/02/28/science-pearl-arctic-research.html) and "Arctic
PEARL tossed away," by Margaret Munro, *Winnipeg Free Press*,
March 24, 2012 www.winnipegfreepress.com/canada/arctic-pearl-
tossed-away-144072336.html
Details on the announcement of CHARS funding (and the Harper quote
from the Cambridge Bay press conference) are drawn from Meagan
Fitzpatrick, "'Science and sovereignty' key to new Arctic research
centre," CBC News, August 23, 2012. www.cbc.ca/news/politics/
story/2012/08/23/pol-harper-thursday-arctic-research.html
The priorities for CHARS cited on page 37 are from the Government of
Canada website. www.science.gc.ca/Canadian_High_Arctic_
Research_Station/CHARS_Priorities-WSE8303E6C-1_En.htm
Details on the restoration of PEARL's funding are drawn from "High Arctic research station saved by new funding," CBC News, May 17, 2013.
www.cbc.ca/news/technology/story/2013/05/17/technology-pearl-
high-arctic-research-station-funding.html
Details on the stealth snowmobile program are drawn from "Operation Silent Snowmobile: New vehicle planned for covert Arctic
ops," Canadian Press, August 21, 2011. www.huffingtonpost.
ca/2011/08/21/operation-silent-snowmobile_n_932334.html
Details on the *Ottawa Citizen*'s reporting on the NRC's snow study are
drawn from "Canadian bureaucracy and a joint study with NASA"
by Tom Spears, *Ottawa Citizen*, April 20, 2012. Documents related
to the Spears query have been archived online at www.scribd.com/
doc/89708162/A-simple-question-a-blizzard-of-bureaucracy
The Stephen Strauss quote on page 41 is drawn from the author's interview with him, April 18, 2012.
Details of the open letter from healthcare professionals to the federal government are drawn from "Eight groups, one message:

restore health coverage for refugees," Patrick Sullivan, Canadian
Medical Association press release, May 24, 2012. www.cma.ca/
eight-groups-one-message-refugees

The Otto Langer quote on page 43 and other details about the DFO's
shortcomings on its Northern Gateway assessment are drawn from
"Northern Gateway review hobbled by budget cuts, critics say," CBC
News/Canadian Press, August 19, 2012. www.cbc.ca/news/canada/
calgary/story/2012/08/19/gateway-pipeline-science.html?cmp=rss

The background and overview of the Experimental Lakes Area here and
in subsequent chapters are drawn from several sources:

Author interview with David Schindler, October 1, 2012.

ELA, "ELA Scientific Milestones and Highlights" (publication of the
ELA, available online at www.experimentallakesarea.ca/images/
ELA%20Scientific%20Milestones%20and%20Highlights.pdf).

Peter Andrey Smith. "Troubled Waters," *The Walrus*, July/August
2013. http://thewalrus.ca/troubled-waters/

Bartley Kives. "Clear thinking Experimental Lakes Area one of most
unusual outdoor labs in the world," *Winnipeg Free Press*, August 17,
2008.

Tom Spears and Ed Struzik. "Scientist's $1,000,000 prize tops
an outspoken career: Freshwater expert David Schindler often
warned by government bosses," *Vancouver Sun*, November 6, 2001.

Stephen Strauss. "Outdoor lakes lab fears extinction," *Globe and Mail*,
June 13, 1996.

Dan Lett. "Lake research revived," *Winnipeg Free Press*, June 10, 1996
(including the "storm of protest" quote).

Jon R. Luoma. "Acid rain studies make real-life labs of Canada lakes,"
New York Times, September 26, 1988.

See also "Phosphorus, detergent, and Canada's Experimental Lakes"
on the Evidence and Error blog, May 20, 2012. http://evidencean-
derror.blogspot.ca/2012/05/phosphorous-detergent-and-canadas.
html

For details on the 2013 study on carcinogens and oil sands pollution, see
"Study provides damning evidence that tar sands development causing carcinogenic pollution in Alberta" by Elizabeth Shope on Switch-
board, the NRDC staff blog, January 8, 2013 (switchboard.nrdc.org/
blogs/eshope/study_provides_damning_evidenc.html). See also
"Lake effect 'smoking gun,'" *Vancouver Sun*, January 26, 2013.

CHAPTER 3: FROM DAWN TO DUSK

The primary source for the discussion of Canada's early scientific history
is *A Curious Field-book: science & society in Canadian history* by Trevor

Harvey Levere and Richard A. Jerrell, Oxford University Press, 1974. Further details about the Royal Canadian Institute's origins and nineteenth-century activities were drawn from the Institute's website (www.royalcanadianinstitute.org).

The discussion of Robert Borden's background and contributions to Canada's progressive tradition is drawn from "Robert Borden and the Rise of the Managerial Prime Minister in Canada" by Ken Rasmussen (paper presented at the 78th Annual Congress of the Social Sciences and Humanities in Ottawa, May 2009; available online at www.cpsa-acsp.ca/papers-2009/Rasmussen.pdf). Further detail was provided by "'Our first duty is to win, at any cost': Sir Robert Borden during the Great War" by Tim Cook, *Journal of Military and Strategic Studies*, *13*(1), Spring 2011.

The section on the emergence of Canadian excellence in government-funded environmental science and the Mulroney-era golden age is based primarily on the author's interviews with David Schindler, John Stone and Jeff Hutchings. (All direct quotes from Schindler and Hutchings in this section are drawn from those interviews unless otherwise noted.) Further detail—including a first-hand account of the 1988 Toronto climate conference—was drawn from "When Canada led the way: a short history of climate change" by Elizabeth May, *Policy Options*, *27*(8), October 2006. Additional detail on the threat to the ELA in 1996 was taken from "Outdoor lakes lab fears extinction" by Stephen Strauss, *Globe and Mail*, June 13,1996, and "Lake research revived" by Dan Lett, *Winnipeg Free Press*, June 10, 1996 (the source of the "storm of protest" quote).

The discussion of the demise of Canada's commercial cod fishery was drawn from a range of sources, including:

Author interview with Jeff Hutchings.

Jeffrey A. Hutchings and Ransom Myers. "What can be learned from the collapse of a renewable resource — Atlantic cod, *Gadus morhua*, of Newfoundland and Labrador," *Canadian Journal of Fisheries and Aquatic Sciences*, *51*(9), 1994.

Jeffrey A. Hutchings, Carl Walters and Richard L. Haedrich. "Is scientific inquiry incompatible with government information control?" *Canadian Journal of Fisheries and Aquatic Sciences*, *54*, 1997.

Jacquelyn Rutherford. "Too many boats chasing too few fish: The collapse of the Atlantic groundfish fishery and the avoidance of future collapses through free market environmentalism," *Studies by Undergraduate Researchers at Guelph*, *2*(1), 2008; available online at https://journal.lib.uoguelph.ca/index.php/surg/article/view/803/1208.

Janet Thomson and Manmeet Ahluwalia. "Remembering the mighty
cod fishery 20 years after moratorium," CBC News, June 29, 2012.
www.cbc.ca/news/canada/story/2012/06/29/f-cod-moratorium-
history.html

CBC Digital Archives. "Cod fishing: 'The biggest layoff in Canadian
history.'" www.cbc.ca/archives/categories/economy-business/
natural-resources/fished-out-the-rise-and-fall-of-the-cod-fishery/
the-biggest-layoff-in-canadian-history.html

Dean Bavington. *Managed Annihilation: An Unnatural History of the
Newfoundland Cod Collapse*, UBC Press, 2010.

Quotes from John Stone on pages 61–62 are from author's interview,
April 11, 2013.

Details on Rona Ambrose's term as Environment minister were drawn
from "Silence of the Lamb" by Jane Taber, *Globe and Mail*, June 2,
2007, and "Harper lowers cone of silence" by Don Martin, *Calgary Herald*, April 13, 2006. The Green Party's eight corrections to
Ambrose's committee testimony are available online at www.green-
party.ca/releases/18.11.2006.

The Stephen Harper quote on page 70 regarding the appointment of
John Baird as Environment minister was taken from "Cabinet shuffle
taps Baird for contentious environment file," CBC News, January 4,
2007. www.cbc.ca/news/canada/story/2007/01/04/cabinet-shuffle.
html

The Baird quotes on pages 70–71 are from "Environment minister
promises tougher climate-change plan," Canwest News Service,
March 18, 2007, and "Canada's Environment Minister Responds
to NRTEE Report," Environment Canada press release, January 7, 2008 (www.ec.gc.ca/default.asp?lang=en&n=714D9AAE-
1&news=0513507F-0361-4840-9ACE-7ACD16125376).

A full recording of Jim Prentice's COP15 speech can be found online at
www.youtube.com/watch?v=m8uRdy-HYSE. Details of the pranks
played on Prentice and his office were drawn from "Environment
Canada hit by 'damn clever' climate stunt," Jane Taber, *Globe and
Mail*, December 14, 2009 (http://www.theglobeandmail.com/news/
politics/ottawa-notebook/environment-canada-hit-by-damn-clever-
climate-stunt/article1346364/) and "Yes Men take credit for fake
climate releases," CBC News, December 14, 2009 (http://www.cbc.
ca/news/canada/story/2009/12/14/hoax-copenhagen-climate.html).
Details of Prentice's advocacy of tougher environmental action before
and after COP15 are drawn from "Jim Prentice, Former Environment
Minister, Pushed Alberta Towards Cap-And-Trade," Huffington
Post/Canadian Press, August 14, 2011 (http://www.huffingtonpost.

ca/2011/08/14/prentice-alberta-cap-trade_n_926365.html) and
"Prentice was ready to curb oilsands: WikiLeaks," CBC News, December 22, 2010 (www.cbc.ca/news/canada/story/2010/12/22/prentice-oil-sands-wikileaks.html). Quotes from Jim Prentice's Calgary speech are drawn from Jason Fekete's article "Prentice tells oilsands to clean up act," *National Post*/Canwest News Service, February 1, 2010 (www.financialpost.com/news-sectors/energy/Prentice+tells+sand s+clean/2509815/story.html). Quotes from Paul Wells's analysis of the speech are taken from "Why Prentice took on the oil sands" by Paul Wells, *Maclean's*, February 5, 2010 (http://www2.macleans. ca/2010/02/05/why-prentice-took-on-the-oil-sands/). The full recording of the "Survivormen: The Wilderness Summit" (special report for CBC's *The National*, October 3, 2010) is online at www.cbc.ca/ thenational/indepthanalysis/story/2010/10/01/national-wilderness-summit.html. The full text of Jim Prentice's resignation is online at www.cbc.ca/news/politics/inside-politics-blog/2010/11/text-of-jim-prentices-statement.html. Quotes from Bruce Cheadle on page 78 are drawn from Bruce Cheadle, "Prentice drops bombshell, quits cabinet," *Winnipeg Free Press*/Canadian Press, November 5, 2011 (www. winnipegfreepress.com/canada/prentice-drops-bombshell-quits-cabi-net-106750723.html).

CHAPTER 4: THE AGE OF WILFUL BLINDNESS

The quote from Environment Canada staff's briefing notes to Michelle Rempel was reported in "Bureaucrats told Stephen Harper's government environmental reforms would be 'very controversial,' records reveal" by Mike De Souza, Postmedia News, January 29, 2013 (http://o.canada.com/2013/01/29/bureaucrats-told-stephen-harpers-government-environmental-reforms-would-be-very-controversial-records-reveal/).

Joe Oliver's "open letter" on opponents of resource extraction was published by numerous outlets, including the *Globe and Mail*, January 9, 2012. www.theglobeandmail.com/news/politics/an-open-letter-from-natural-resources-minister-joe-oliver/article4085663/

Details of oil-and-gas industry consultation and input on Bill C-38 is drawn from "Energy industry letter suggested environmental law changes," by Max Paris, CBC News, January 9, 2013 (www.cbc.ca/news/politics/story/2013/01/09/pol-oil-gas-industry-letter-to-government-on-environmental-laws.html); "Pipeline industry pushed changes to Navigable Waters Protection Act: documents" by Heather Scoffield, Canadian Press, February 20, 2013 (http://globalnews.ca/

news/395183/pipeline-industry-pushed-changes-to-navigable-waters-protection-act-documents-5/); and "Stephen Harper's 'omnibus' strategy to overhaul green laws was proposed by oil industry, says records" by Mike De Souza, Postmedia News, April 10, 2013.

The quote from Rick Smith of Environmental Defence on page 86 is taken from "Canadian government overhauling environmental rules to aid oil extraction" by Juliet Eilperin, *Washington Post*, June 3, 2012 (http://articles.washingtonpost.com/2012-06-03/national/35460476_1_tar-sands-canadian-government-macdonald-laurier-institute). All other Smith quotes in this section and paraphrases of his interpretation of the shift in tone before and after Oliver's letter are from the author's interview with Smith, April 24, 2012.

The quote from Chantal Hébert on page 88 is taken from Hébert's column "Tories scrambling to keep energy-based economic agenda on track: Hébert," *Toronto Star*, March 21, 2013. www.thestar.com/news/canada/2013/03/21/tories_scrambling_to_keep_energybased_economic_agenda_on_track_hbert.html

Details about Natural Resources Canada briefing notes to Joe Oliver are taken from "Oil sands 'landlocked' due to environmental concerns and market bottlenecks" by Mike De Souza, December 7, 2011. http://business.financialpost.com/2012/07/11/oil-sands-land-locked-due-to-environmental-concerns-and-market-bottlenecks/?_lsa=1eb2-3fc6

Details about Environment Canada staff's advice to Peter Kent are taken from "Environment Canada offers Peter Kent tips to describe impact of climate change" by Mike De Souza, Postmedia News, September 28, 2012. http://o.canada.com/2012/09/28/environment-canada-offers-peter-kent-tips-on-making-useful-comments-about-climate/

Ignored trans-fat recommendations were reported in "Feds drop trans-fat monitoring in foods, despite expert advice" by Sarah Schmidt, Postmedia News, July 20, 2012. www.canada.com/health/Feds+drop+trans+monitoring+foods+despite+expert+advice/6960561/story.html

The quote from Peter Kent on page 90 is taken from "Independent analysis further eroded with closing of NRTEE" by Daniel Veniez, iPolitics.ca, April 2, 2012 (www.ipolitics.ca/2012/04/02/dan-veniez-independent-analysis-further-eroded-with-closing-of-nrtee/). Kent's lockdown of the NRTEE's files is reported in "Peter Kent orders doomed advisory panel to turn over website files" by Mike De Souza, March 26, 2013. http://o.canada.com/2013/03/26/peter-kent-orders-doomed-advisory-panel-to-turn-over-website-files/

The Robert Sopuck quote on page 90 is taken from "Harper government's muzzling of scientists a mark of shame for Canada" by Jeffrey Hutchings, March 15, 2013. www.thestar.com/opinion/commentary/2013/03/15/harper_governments_muzzling_of_scientists_a_mark_of_shame_for_canada.html

The Joseph Heath quotes on pages 90–91 are from Heath's column "In defence of sociology," *Ottawa Citizen*, April 30, 2013. www.ottawacitizen.com/opinion/op-ed/defence+sociology/8317722/story.html

Details of MP David Wilks's dissenting YouTube video are drawn from "Re-education of David Wilks a lesson in the decline of Parliament" by Andrew Coyne, *National Post,* May 30, 2012. http://fullcomment.nationalpost.com/2012/05/30/andrew-coyne-re-education-of-david-wilks-a-lesson-in-the-decline-of-parliament/

The federal cabinet's snub of the IPCC Nobel laureates was reported in "Harper government absent from ceremony for climate change experts," Canwest News Service, February 13, 2008. www.canada.com/topics/news/politics/story.html?id=29fb149b-a842-4f5e-a4d0-5c6be778df66&k=63539

Details of the Conservative government's use of "job-killing carbon tax" as a rhetorical attack are drawn from "Tory carbon-tax campaign against NDP frames debate, tough to counteract," Huffington Post/Canadian Press, September 19, 2012 (www.huffingtonpost.ca/2012/09/19/tory-carbon-tax-campaign-_n_1898231.html) and "Great moments in farce: the definitive collection" by Aaron Wherry, *Maclean's*, October 15, 2012 (www2.macleans.ca/2012/10/15/great-moments-in-farce-the-definitive-collection/). Wherry's "Great moments in farce," *Maclean's*, October 15, 2012 (www2.macleans.ca/2012/10/15/great-moments-in-farce-the-definitive-collection/) and "John Baird put a price on carbon in writing and signed his name to it," *Maclean's*, October 23, 2012 (www2.macleans.ca/2012/10/23/john-baird-put-a-price-on-carbon-in-writing-and-signed-his-name-to-it/) also detail Baird and Prentice's rhetorical commitments to a price on carbon. The Wherry quote on pages 94–95 is from "The Commons: The joke is on you, Canada," *Maclean's*, September 17, 2012 (www2.macleans.ca/2012/09/17/the-commons-the-joke-is-on-you-canada/).

The Michael Den Tandt quote on page 96 is from "If Keystone goes awry, Conservatives will have only themselves to blame," Postmedia News, February 21, 2013. http://o.canada.com/2013/02/21/0222-coldentandt/#.USaZflqsbWM

Figures on Economic Action Plan costs were taken from "Economic Action Plan Ads: Harper Cites Pride To Defend $113 Million In Ads"

by Bruce Cheadle, Huffington Post/Canadian Press, May 7, 2013.
www.huffingtonpost.ca/2013/05/07/economic-action-plan-ads-
harper_n_3232543.html

Details of the government's withdrawal from the UN desertification convention are taken from "Conservatives defend withdrawal from UN
drought 'talkfest,'" CBC News, March 28, 2013. www.cbc.ca/news/
politics/story/2013/03/28/pol-un-convention-drought-desertification-
harper-baird.html

The quote from Peter Kent's anniversary statement about the Montreal Protocol is taken from "Canada Celebrates 25 Years of Success
with Montreal Protocol," Environment Canada press release, September 14, 2012. www.ec.gc.ca/default.asp?lang=En&n=FFE36B6D-
1&news=AE9117A6-E9D4-43DB-A6DE-6E779B550E1E

The full text of Leona Aglukkaq's speech at the International Polar Year
conference on April 26, 2012, can be found at www.hc-sc.gc.ca/ahc-
asc/minist/speeches-discours/_2012/2012_04_26-eng.php

Details of Ducks Unlimited's foreign donations are taken from "Environmental charities not biggest recipients of foreign cash, tax returns
show," Canadian Press, May 10, 2012.

Details about the reduced capacity of government agencies to respond
to environmental problems are drawn from several sources:
Gloria Galloway. "Cuts at Environment Canada mean fewer
left to clean up oil-spill mess," *Globe and Mail*, April 13, 2012.
www.theglobeandmail.com/news/politics/cuts-at-environ-
ment-canada-mean-fewer-left-to-clean-up-oil-spill-mess/
article4178488/

"Ottawa axes ocean pollution monitoring program," *Victoria Times
Colonist*, May 23, 2012.

Peter Ross. "Canada's mass firing of ocean scientists brings 'silent
summer,'" *Environmental Health News*. www.environmentalhealth
news.org/ehs/news/2012/opinion-mass-firing-of-canada2019s-
ocean-scientists

Jonathan Gatehouse. "When science goes silent," *Maclean's*, May 3,
2013. www2.macleans.ca/2013/05/03/when-science-goes-silent/

Margaret Munro. "Closure of fisheries' libraries called a 'disaster' for science," Postmedia News, April 14, 2013. www.canada.
com/technology/Closure+fisheries+libraries+called+disaster+scie
nce/8241123/story.html

CBC News. "DFO gets F in free expression from journalism group,"
May 2, 2013. www.cbc.ca/news/technology/story/2013/05/02/
science-free-expression-report-card-dfo-cjfe.html

The Otto Langer quote on page 104 is from "Northern Gateway
review hobbled by budget cuts, critics say," cbc News/Cana-
dian Press, August 19, 2012. www.cbc.ca/news/canada/calgary/
story/2012/08/19/gateway-pipeline-science.html?cmp=rss

For details on outdoor organizations' opposition to Fisheries Act
changes, see "Outdoor groups worried that Federal budget may 'gut'
Fisheries Act," Scott Gardner, *Outdoor Canada*. http://outdoorcanada.
ca/19417/news/articles/outdoor-groups-worried-that-federal-budget-
may-%E2%80%9Cgut%E2%80%9D-fisheries-act?rel=author

The Conrad Fennema quote on page 104 is taken from his letter to the
government, posted online at www.afga.org/news/article/Potential-
amendments-to-the-Fisheries-Act/8bc6e880efdddbf48e1dc09d27df
f50a/

The Jeff Hutchings quote on page 105 is drawn from "Canadian scientists slam weakening of federal Fisheries Act," Peter O'Neil and Larry
Pynn, *Vancouver Sun*, May 28, 2012. www.vancouversun.com/Canadi
an+scientists+slam+weakening+federal+Fisheries/6691159/story.html

The section on anti-scientific beliefs and American democracy's origins
is drawn from "Antiscience Beliefs Jeopardize U.S. Democracy"
by Shawn Lawrence Otto, *Scientific American*, October 16, 2012.
www.scientificamerican.com/article.cfm?id=antiscience-beliefs-
jeopardize-us-democracy

The full text of Immanuel Kant's essay "What is Enlightenment?" is available online at www.columbia.edu/acis/ets/ccread/etscc/kant.html

The primary source for the section discussing changes to the National
Research Council is an unnamed research scientist whose identity is
being protected by the author. In addition to an extensive telephone
interview, the nrc source provided the text of numerous emails sent
by John McDougall to all staff detailing the new strategies to be
implemented at the nrc. Additional detail is taken from:

cbc News. "Scientists still wary after science minister says he believes
in evolution," March 18, 2009. www.cbc.ca/news/technology/
story/2009/03/18/tech-090318-gary-goodyear-evolution-
scientists.html

cbc News. "National Research Council to 'refocus' to serve business,"
March 6, 2012, from which the Gary Goodyear quote on page 111
is taken. www.cbc.ca/news/technology/story/2012/03/06/tech-
nology-goodyear-national-research-council.html

"nrc staff enraged by gift cards," *Winnipeg Free Press*, July 5, 2012.
www.winnipegfreepress.com/local/nrc-staff-enraged-by-gift-
cards-161407515.html

Tom Spears. "Pure science research drops sharply at National Research Council," *Ottawa Citizen*, May 7, 2013. www.ottawacitizen.com/technology/Pure+science+research+drops+sharply+Natio nal+Research+Council/8351165/story.html#ixzz2xY0JVSR8

Kate Allen. "National Research Council 'open for business,' Conservative government says," *Toronto Star*, May 7, 2013, from which the 2013 Goodyear quotes on page 113 are drawn. www.thestar.com/ news/canada/2013/05/07/national_research_council_open_ for_business_conservative_government_says.html

The quote from David Schindler on page 113 is from the author's interview with him.

Details about Bill Buxton's experiences at the NRC and in the computer graphics industry are drawn primarily from Buxton's own writing on the subject:

"The Cost of Saving Money: The Folly of Research Funding Policy in Canada," *Research Money*, March 2001. http://billbuxton.com/ ResearchFunding.html

"My Vision Isn't *My* Vision," from *HCI Remixed* (MIT Press, 2008; www.billbuxton.com/MyVision.pdf)

"The Long Nose of Innovation," *Business Week*, January 2, 2008, from which the quote on pages 116–17 is taken. www.businessweek. com/stories/2008-01-02/the-long-nose-of-innovationbusiness-week-business-news-stock-market-and-financial-advice

CHAPTER 5: LOST IN THE DARK

Details about the leaking Environment Canada facility are drawn from "Environment Canada sent itself a warning after diesel leak at water research building" by Mike De Souza, Postmedia News, January 21, 2013. http://o.canada.com/2013/01/21/environment-canada-sent-itself-a-warning-after-diesel-leak-at-water-research-building/

Details about new DFO confidentiality rules are drawn from "Scientist calls new confidentiality rules on Arctic project 'chilling'" by Margaret Munro, Postmedia News, February 13, 2013. http://o.canada.com/ 2013/02/13/feds-new-confidentiality-rules-on-arctic-project-called-chilling/

The Muenchow blog post quoted on page 120 was reprinted in "Canadian federal research deal 'potentially muzzles' U.S. scientists," CBC News, February 16, 2013. www.cbc.ca/news/technology/ story/2013/02/15/science-audio-munchow-scientist-muzzling.html

Details about Parks Canada's experiences with the PCO are drawn from "Harper's communications unit bigfoots Parks Canada news conference," *Maclean's*/Canadian Press, March 17, 2013. www2.macleans.

ca/2013/03/17/harpers-communications-unit-bigfoots-parks-canada-news-conference/

The transfer of ELA management to Ontario's government and the IISD is reported in "Experimental Lakes Area research facility in Ontario finds new manager," *Toronto Star*, May 9, 2013. www.thestar.com/news/canada/2013/05/09/experimental_lakes_area_research_facility_in_ontario_finds_new_manager.html

The MyKawartha.com article on Dean Del Mastro's support for the new deal is "MP Del Mastro supports Environmental Lakes Area deal," May 13, 2013. www.mykawartha.com/community-story/3716789-mp-del-mastro-supports-environmental-lakes-area-deal/

The Jules Blais quote on pages 122–23 is drawn from "Experimental Lakes Area in danger of closing," CBC News, March 7, 2013. www.cbc.ca/news/canada/story/2013/03/07/pol-experimental-lakes-to-be-mothballed.html

Details on the Bamfield closure are drawn from "Science Cuts: Ottawa Views Pure Science As 'Cash Cow,' Critics Say," Huffington Post, May 7, 2013 (www.huffingtonpost.ca/2013/05/07/science-cuts-canada_n_3228151.html), and "Four scientists tally the cost of science funding cuts," *Toronto Star*, May 19, 2013. www.thestar.com/opinion/commentary/2013/05/19/four_scientists_tally_the_cost_of_science_funding_cuts.html

The details of Pat Sutherland's dismissal and quote from Andrew Gregg are drawn from "Cold comfort" by Don Butler, *Ottawa Citizen*, November 22, 2012.

ACKNOWLEDGEMENTS

T HE GENESIS of *The War on Science* was an assign-
ment for *Corporate Knights* magazine, so my first
debt of gratitude is to my editor there, Tyler Ham-
ilton, for directing my attention to Canada's muzzled scientists and this government's wilful blindness.

That seed was germinated by Nancy Flight, this book's
steadfast editor, who steered it through tumultuous waters
of a near-bankruptcy and other publishing-business chaos,
but remained keen-eyed, cheerful, and enthusiastic throughout. My thanks as well to Lesley Cameron for her careful
copy edits, and to Rob Sanders, Shirarose Wilensky and Zoe
Grams at Greystone for their support.

I am also deeply indebted to dozens of hard-working journalists who have been covering the Conservative government's activities daily. I want to extend my special thanks to
two in particular: Mike De Souza and Margaret Munro at
Postmedia News. De Souza's dogged reporting on environmental issues in the Harper years—and especially his Access

to Information requests for Environment Canada files—provided reams of vital primary research to this book. Munro has been equally meticulous in her reporting on the science beat, especially on the muzzling of government scientists. Their work highlights the irreplaceable public service that professional beat reporting provides in a democracy.

In addition to the sources cited by name in the text, a great many colleagues lent expertise and enthusiasm to this book (and/or assisted me in an inspiring run at the parliamentary seat in Calgary Centre in the midst of the research). My huge and everlasting thanks to Nancy Close, Cheri Macaulay, Leor Rotchild, Natalie Odd, Marc Doll, Nimra Amjad-Archer, Kurt Archer, A.J. Brouillet, Chantal Chagnon, Jodi Christensen, Kris Demeanor, Janice Dixon, Dale D'Silva, Shane Gallagher, Jamie Herington, Gillian Hickie, Irene Johansen, Cheryl Johnson, Jana Johnson, John Manzo, Kate McKenzie, Alex Middleton, Peter Oliver, Evan Osenton, Becky Rock, Nicole Schon, Peter Schryvers, Natalie Sit, David Suzuki, Dave Thompson, Chris Wharton, Sheri-D Wilson, Dave Bagler, Katie Gibbs, and the hundreds of other Turner4YYC volunteers and thousands of Calgary voters who joined me in the most inspirational political project of my life so far. And my special thanks to Elizabeth May, who threw me into the maelstrom, and in so doing restored my faith in parliamentary democracy.

My thanks as well to Richard Peltier, Boris Worm, Andrew Weaver, Craig Pyette, Christopher Frey, Trena White, Janice Paskey, Jeremy Van Loon, John Vaillant, Gian-Carlo Carra, Trevor Day, Marcello DiCintio, Gerald Butts, Jay Ingram, Anne Casselman, Brian Singh, Dan Woynillowicz, Meike Wielebski, Angus MacIssac, Terry Rock, Ed Whittingham, Andrew Heintzman, Justin Trudeau, Alex Steffen, Jeremy

Klaszus, John Streicker, Naheed Nenshi, Gillian Deacon, Grant Gordon, and Jessica Cameron for their advice, advocacy, friendship, hospitality, and professional support.

I am deeply indebted to the incomparable Berton House in Dawson City, Yukon, for providing a comfortable and endlessly fascinating retreat for the writing of a portion of this book. My thanks to the Writers' Trust of Canada, and especially James Davies for administering the writer-in-residence program there, to the Canada Council for the Arts for providing funding support, and to Dawson's vibrant, enthusiastic, welcoming arts community for enabling an unforgettable experience. My particular thanks go out to these amazing Dawsonites: Lulu Keating, Dan and Laurie Sokolowski, Shelley and Greg Hakonson, Gord Macrae, Meg Walker, and Peter Menzies.

As ever, I am thankful for more support on the home front than I could possibly properly acknowledge here. My parents, John and Margo Turner, and my father-in-law, Bruce Bristowe, were their usual (and extraordinary) steadfast selves throughout. My children, Sloane and Alexander, continue to inspire me daily to fight for this country's brightest possible future. And in this and all projects, my deepest gratitude and most substantial debt is to my wife, Ashley Bristowe, who remains my best editor and most thoughtful collaborator, and without whom this book and much else would simply not be possible.

INDEX